QUANTUM COMPUTING

COMPUTER SCIENCE, PHYSICS, AND MATHEMATICS

4 BOOKS IN 1

BOOK 1
QUANTUM COMPUTING DEMYSTIFIED: A BEGINNER'S GUIDE

BOOK 2
MASTERING QUANTUM COMPUTING: A COMPREHENSIVE GUIDE FOR INTERMEDIATE LEARNERS

BOOK 3
ADVANCED QUANTUM COMPUTING: EXPLORING THE FRONTIERS OF COMPUTER SCIENCE, PHYSICS, AND MATHEMATICS

BOOK 4
QUANTUM COMPUTING: A MULTIDISCIPLINARY APPROACH FOR EXPERTS

ROB BOTWRIGHT

Published by Rob Botwright
Library of Congress Cataloging-in-Publication Data
ISBN 978-1-83938-632-9
Cover design by Rizzo

Disclaimer

The contents of this book are based on extensive research and the best available historical sources. However, the author and publisher make no claims, promises, or guarantees about the accuracy, completeness, or adequacy of the information contained herein. The information in this book is provided on an "as is" basis, and the author and publisher disclaim any and all liability for any errors, omissions, or inaccuracies in the information or for any actions taken in reliance on such information. The opinions and views expressed in this book are those of the author and do not necessarily reflect the official policy or position of any organization or individual mentioned in this book. Any reference to specific people, places, or events is intended only to provide historical context and is not intended to defame or malign any group, individual, or entity. The information in this book is intended for educational and entertainment purposes only. It is not intended to be a substitute for professional advice or judgment. Readers are encouraged to conduct their own research and to seek professional advice where appropriate. Every effort has been made to obtain necessary permissions and acknowledgments for all images and other copyrighted material used in this book. Any errors or omissions in this regard are unintentional, and the author and publisher will correct them in future editions.

BOOK 1 - QUANTUM COMPUTING DEMYSTIFIED: A BEGINNER'S GUIDE

BOOK 2 - MASTERING QUANTUM COMPUTING: A COMPREHENSIVE GUIDE FOR INTERMEDIATE LEARNERS

BOOK 3 - ADVANCED QUANTUM COMPUTING: EXPLORING THE FRONTIERS OF COMPUTER SCIENCE, PHYSICS, AND MATHEMATICS

BOOK 4 - QUANTUM COMPUTING: A MULTIDISCIPLINARY APPROACH FOR EXPERTS

Introduction

Welcome to the world of quantum computing—a revolutionary field that marries the profound principles of quantum mechanics with the computational power to transform our digital landscape. In this book bundle, "Quantum Computing: Computer Science, Physics, and Mathematics," we embark on a journey through the multifaceted world of quantum computing, offering a comprehensive exploration from beginner to expert levels.

Book 1, "Quantum Computing Demystified: A Beginner's Guide," serves as your portal into the quantum realm. Here, we unravel the enigmatic concepts of quantum mechanics and quantum computing, making them accessible to those taking their first steps in this captivating field. From qubits and superposition to quantum algorithms, this book lays a solid foundation for your quantum voyage.

In Book 2, "Mastering Quantum Computing: A Comprehensive Guide for Intermediate Learners," we venture deeper into the quantum landscape. Intermediate learners will find a wealth of knowledge here, as we delve into advanced topics, quantum programming, and algorithm design. With hands-on examples and in-depth explanations, you'll gain the skills needed to harness quantum power effectively.

Book 3, "Advanced Quantum Computing: Exploring the Frontiers of Computer Science, Physics, and Mathematics," propels us to the forefront of quantum computing's cutting-edge. Dive into quantum error correction, quantum *cryptography, and quantum simulations. Explore the complex challenges and captivating possibilities that await at the vanguard of this transformative technology.*

Lastly, in Book 4, "Quantum Computing: A Multidisciplinary Approach for Experts," we transcend boundaries and discover that quantum computing is not confined to one discipline—it's a bridge connecting computer science, physics, mathematics, and beyond. Explore the multifaceted applications of quantum computing in various domains, recognizing its potential to reshape industries and address global challenges.

Together, these four books offer a comprehensive and multidimensional view of quantum computing. Whether you're a curious beginner or a seasoned expert, this bundle is your gateway to understanding the quantum world. From the fundamental principles to the frontiers of research, we invite you to embark on a transformative journey through the realm of quantum computing.

BOOK 1
QUANTUM COMPUTING DEMYSTIFIED
A BEGINNER'S GUIDE
ROB BOTWRIGHT

Chapter 1: Understanding the Basics of Quantum Mechanics

Quantum mechanics, a branch of physics that delves into the fundamental properties of the smallest particles in the universe, has revolutionized our understanding of the natural world. It has revealed a reality that is far more mysterious and counterintuitive than classical physics could ever have imagined. At its core, quantum mechanics explores the behavior of particles at the quantum level, where the classical rules of physics no longer apply. Instead, we encounter phenomena that challenge our intuition and force us to rethink the very nature of reality.

One of the key principles of quantum mechanics is the idea that particles can exist in a superposition of states. This means that, unlike classical objects that have definite properties such as position and velocity, quantum particles can exist in multiple states simultaneously. This concept was famously illustrated by Erwin Schrödinger's thought experiment involving a cat that could be both alive and dead at the same time, depending on the quantum state of a radioactive atom.

Superposition is not limited to just the cat in Schrödinger's scenario; it applies to all quantum particles. For example, an electron can be in a superposition of different energy levels in an atom, and a photon of light can exist in a superposition of different polarizations. This fundamental property of particles forms the foundation of quantum computing, where quantum bits or qubits can represent both 0 and 1 at the same time, allowing for the potential of exponentially faster calculations.

Another fascinating aspect of quantum mechanics is the uncertainty principle, first formulated by Werner Heisenberg. This principle asserts that there is a fundamental limit to how precisely we can simultaneously know certain pairs of properties of a particle, such as its position and momentum. The more accurately we try to measure one of these properties, the less accurately we can know the other. This inherent uncertainty at the quantum level challenges our classical notion of determinism, where we could, in principle, predict the future with absolute certainty if we knew all the initial conditions.

Furthermore, quantum mechanics introduces the concept of entanglement, a phenomenon Albert Einstein famously referred to as "spooky action at a distance." When two particles become entangled, their quantum states become correlated in such a way that the measurement of one particle instantaneously affects the state of the other, even if they are separated by vast distances. This eerie interconnectedness defies classical notions of locality and has led to numerous experiments and debates about the nature of quantum reality.

As we delve deeper into the quantum world, we encounter wave-particle duality, a central tenet of quantum mechanics. This principle suggests that particles, such as electrons and photons, can exhibit both wave-like and particle-like properties depending on how they are observed. This duality is exemplified in the famous double-slit experiment, where particles sent through two slits create an interference pattern on a screen, as if they were behaving like waves. However, when we observe which slit the particle goes through, it behaves as a discrete particle, not a wave.

Wave-particle duality underscores the probabilistic nature of quantum mechanics. Instead of definite trajectories, quantum particles are described by wavefunctions that represent the probability of finding a particle in a particular state. These wavefunctions evolve over time according to the Schrödinger equation, which governs the dynamics of quantum systems. The mathematical formalism of quantum mechanics is elegant and powerful, allowing us to make accurate predictions about the behavior of particles at the quantum level.

In addition to wave-particle duality, the concept of quantum tunneling highlights another intriguing aspect of quantum mechanics. Quantum tunneling occurs when a particle penetrates a potential energy barrier that classical physics would deem impenetrable. This phenomenon is essential for the operation of devices such as tunnel diodes and plays a significant role in nuclear fusion processes inside stars.

Understanding the quantum nature of particles requires a shift in perspective from classical intuition to embracing the probabilistic and counterintuitive aspects of the quantum realm. It challenges our perception of reality and forces us to accept that, at the quantum level, particles exist in a state of constant flux, transitioning between different possibilities until they are observed.

Quantum mechanics has not only expanded our understanding of the microscopic world but has also led to the development of groundbreaking technologies, such as lasers, transistors, and MRI machines, which rely on quantum principles. Moreover, quantum computing, with its potential to revolutionize fields like cryptography and optimization, is on the horizon, promising to take advantage of the quantum nature of particles to solve complex problems that are beyond the reach of classical computers.

In summary, the quantum nature of particles is a fundamental aspect of modern physics that challenges our classical worldview. It introduces concepts such as superposition, uncertainty, entanglement, wave-particle *duality, and quantum tunneling, which collectively paint a rich and enigmatic picture of the quantum world. Embracing these principles has not only deepened our understanding of nature but has also paved the way for transformative technologies that will shape the future of science and technology.*

In the realm of quantum mechanics, one of the most perplexing and intriguing phenomena is wave-particle duality, which challenges our classical understanding of the nature of particles. It suggests that particles, such as electrons and photons, can exhibit both wave-like and particle-like properties depending on how they are observed or measured. This dual nature was first introduced by Louis de Broglie in the early 20th century and later solidified by experiments such as the double-slit experiment.

In the double-slit experiment, a beam of particles, often electrons or photons, is directed at a barrier with two narrow slits. What makes this experiment particularly intriguing is that when these particles pass through the slits and strike a screen on the other side, they create an interference pattern, similar to the pattern produced by waves in water. This outcome is unexpected if particles were purely particles, as classical physics would suggest.

However, the mystery deepens when scientists decide to monitor which slit the particles pass through. When they do this, the interference pattern disappears, and the particles behave more like discrete particles, landing on the screen in two distinct lines corresponding to the positions of the slits. It's as if the act of measurement collapses the wave-like behavior into a particle-like behavior.

This phenomenon is not limited to the double-slit experiment alone. It extends to various aspects of quantum mechanics and has profound implications for our understanding of the quantum world. Wave-particle duality suggests that particles are not confined to classical trajectories with definite positions and velocities but exist in a state of probabilistic flux, described by a wavefunction that represents the probability of finding a particle at a particular location.

The wavefunction itself evolves over time according to the Schrödinger equation, a fundamental equation in quantum mechanics. The wave-like aspect of particles is evident in the behavior of these wavefunctions, which can exhibit interference patterns and diffraction effects similar to those observed in classical wave phenomena.

Yet, when we make a measurement or an observation, the wavefunction collapses to a single definite value, determining the outcome of the measurement. This transition from a probabilistic wave-like state to a definite particle-like state upon observation is a central tenet of quantum mechanics and wave-particle duality.

The implications of wave-particle duality are profound and far-reaching. They challenge our classical intuition and demand a shift in perspective when dealing with particles at the quantum level. Unlike classical objects with well-defined properties, quantum particles exist in a state of uncertainty until observed. This inherent uncertainty is encapsulated in Heisenberg's Uncertainty Principle, which states that there is a fundamental limit to how precisely we can simultaneously know certain pairs of properties of a particle, such as its position and momentum.

Furthermore, the wave-particle duality has significant consequences for the development of quantum technologies and our understanding of the behavior of matter and energy at the smallest scales. It underlies the operation of devices such as electron microscopes and diffraction gratings, which rely on the wave-like behavior of particles to achieve high-resolution imaging and precise measurements.

In the realm of quantum computing, wave-particle duality plays a pivotal role. Quantum bits or qubits, which can represent both 0 and 1 simultaneously, rely on the probabilistic and wave-like nature of particles to perform quantum computations. This ability to harness superposition and interference, two key characteristics of wave-particle duality, holds the promise of exponentially faster calculations in certain applications.

As we delve deeper into the quantum world, it becomes increasingly apparent that wave-particle duality is not an anomaly but a fundamental feature of nature at the quantum level. It forces us to reevaluate our classical intuitions and embrace the probabilistic and dynamic nature of particles in this mysterious realm.

In summary, wave-particle duality is a cornerstone of quantum mechanics, revealing that particles can exhibit both wave-like and particle-like properties depending on how they are observed or measured. This duality challenges classical notions of determinism and definiteness, introducing a probabilistic and uncertain aspect to the behavior of particles. It has profound implications for our understanding of the quantum world and plays a central role in the development of quantum technologies that have the potential to revolutionize computing and other scientific endeavors.

Chapter 2: Quantum Bits and Qubits

In the ever-evolving landscape of computing, the distinction between classical and quantum bits forms a foundational divide. Classical bits, the bedrock of traditional digital computing, represent the smallest unit of information as either a 0 or a 1. In contrast, quantum bits, often referred to as qubits, introduce a realm of possibilities beyond the binary world of classical computing. These two forms of bits serve as the basis for distinct computing paradigms, each with its unique capabilities and potential.

At the heart of classical computing is the concept of bits, which can exist in one of two states: 0 or 1. This binary representation forms the foundation of digital information processing, allowing classical computers to execute algorithms and solve problems by manipulating strings of these binary digits. It's a deterministic world where each bit is unambiguously either 0 or 1, enabling precise calculations and logical operations.

However, quantum bits, or qubits, depart from this classical simplicity. Qubits, unlike classical bits, exploit the principles of quantum mechanics to represent information in a more versatile way. A qubit can exist in a superposition of states, meaning it can be simultaneously both 0 and 1. This superposition property opens up a realm of possibilities, as qubits can process multiple pieces of information at once, exponentially increasing their computational power compared to classical bits.

In addition to superposition, qubits possess another crucial characteristic known as entanglement. When qubits become entangled, their quantum states become correlated in such a way that the measurement of one qubit instantaneously affects the state of the other, even if they are separated by vast distances. This phenomenon, described by Albert Einstein as "spooky action at a distance," enables qubits to be interconnected in ways that classical bits cannot.

The unique properties of qubits give rise to quantum computing, a field that has garnered significant attention and excitement in recent years. Quantum computers, leveraging the power of qubits, have the potential to tackle complex problems that would be infeasible for classical computers to solve in a reasonable amount of time. Tasks such as simulating quantum systems, optimizing complex processes, and breaking certain cryptographic codes could be revolutionized by quantum computing.

One of the defining characteristics of quantum computing is its ability to perform quantum parallelism. Classical computers solve problems sequentially, one step at a time, while quantum computers can explore multiple possibilities simultaneously due to the superposition property of qubits. This means that quantum algorithms can, in some cases, provide exponential speedup over their classical counterparts.

Shor's algorithm, for example, demonstrates the power of quantum parallelism by efficiently factoring large numbers, a task that underpins the security of many encryption methods. Shor's algorithm can factor large numbers exponentially faster than the best-known classical algorithms, posing a potential threat to classical encryption systems.

Grover's algorithm, on the other hand, highlights the advantage of quantum computing in search problems. It can search an unsorted database of items in roughly the square root of the number of steps required by classical algorithms. Grover's algorithm has implications for applications such as database search, optimization, and even artificial intelligence.

Despite these remarkable capabilities, quantum computing is not a panacea for all computational problems. Quantum algorithms excel in specific domains but may not offer advantages for tasks that are inherently sequential or do not take advantage of the quantum parallelism inherent in qubits. Furthermore, building and maintaining stable qubits and maintaining their quantum coherence remain formidable technical challenges.

Quantum computing also brings a new level of complexity to the field of error correction. The delicate nature of qubits makes them susceptible to noise and errors from their environment, leading to the need for error-correcting codes and fault-tolerant quantum computing architectures.

In the pursuit of practical quantum computers, several approaches have emerged, including superconducting qubits, trapped ion qubits, topological qubits, and others. Each approach offers unique advantages and faces its own set of technical challenges, such as maintaining qubit coherence and scalability.

In the realm of quantum communication, qubits also offer intriguing possibilities. Quantum key distribution, for instance, allows for secure communication based on the principles of quantum entanglement and the Heisenberg Uncertainty Principle. This method can provide a level of security that is theoretically unbreakable, making it appealing for applications where data privacy is critical.

In summary, the distinction between classical bits and quantum bits represents a fundamental divide in the world of computing. Classical bits are binary, deterministic, and serve as the basis for traditional digital computing, while qubits, with their superposition and entanglement properties, introduce a new paradigm of quantum computing. Quantum *computing has the potential to revolutionize fields such as cryptography, optimization, and quantum simulation but also poses significant technical challenges. As we continue to explore and harness the power of qubits, the boundary between classical and quantum computing remains an exciting frontier in the world of technology and science.*

In the fascinating realm of quantum mechanics, the concept of superposition and its intricate relationship with measurement stand as fundamental principles that underpin the quantum world. Superposition, a cornerstone of quantum theory, defies classical intuition by allowing quantum particles to exist in multiple states simultaneously. This phenomenon challenges our conventional understanding of how physical systems behave and forms the basis for many quantum technologies, including quantum computing.

Superposition can be thought of as a blending of possibilities. Imagine a quantum bit, or qubit, which is the quantum counterpart of a classical bit. While a classical bit can represent either a 0 or a 1, a qubit can exist in a superposition of both 0 and 1 states simultaneously. This means that, until measured, the qubit is in a probabilistic state that encompasses both possibilities. It's as if the qubit is exploring multiple realities at once.

Mathematically, superposition is described by a complex number called a probability amplitude. The square of the absolute value of this amplitude determines the probability of measuring the qubit in a particular state. This probabilistic nature of quantum systems is one of the defining characteristics of the quantum world.

To illustrate the concept of superposition, consider the famous example of Schrödinger's cat. In this thought experiment, a cat is placed in a sealed box with a radioactive atom. If the atom decays, a device inside the box is triggered to release poison, which kills the cat. However, according to quantum mechanics, until we open the box and make an observation, the atom is in a superposition of decayed and undecayed states, and the cat is in a superposition of dead and alive states. It's only when we open the box and measure the state of the atom that we determine the fate of the cat.

Superposition is not limited to qubits or macroscopic scenarios like Schrödinger's cat. It applies to all quantum particles, from electrons and photons to atoms and molecules. For instance, an electron in an atom can exist in a superposition of energy levels, which gives rise to the rich spectra observed in atomic physics. The ability of particles to exist in superpositions of states has profound implications for our understanding of matter and energy at the quantum level.

Quantum computers harness the power of superposition to perform certain calculations exponentially faster than classical computers. Instead of evaluating possibilities sequentially, a quantum algorithm can explore multiple solutions simultaneously. For example, Shor's algorithm, which can factor large numbers efficiently, exploits the superposition property to break classical encryption methods. Grover's algorithm, used for database search and optimization, relies on superposition to speed up the search process.

However, the story of quantum superposition is incomplete without discussing measurement. Measurement is a fundamental aspect of quantum mechanics, and it plays a unique role in determining the outcome of quantum systems. When we measure a quantum system, it collapses from its superposition of states into a single definite state, as described by the famous collapse of the wavefunction.

The collapse of the wavefunction is a profound and mysterious aspect of quantum mechanics. It means that *before measurement, particles exist in a state of ambiguity, but as soon as we observe them, they assume a particular value with certainty. This transition from potentiality to actuality upon measurement has been a topic of philosophical debate and exploration for decades.*

Moreover, the act of measurement itself can influence the outcome. The Heisenberg Uncertainty Principle, another fundamental principle of quantum mechanics, asserts that there is a limit to how precisely we can simultaneously know certain pairs of properties of a particle, such as its position and momentum. When we make a measurement to determine one property with high precision, the uncertainty in the other property increases.

This inherent uncertainty poses a challenge for scientists and engineers working with quantum systems. The delicate nature of quantum states and their susceptibility to measurement-induced disturbance necessitate careful consideration and control in quantum experiments and technologies.

In summary, superposition and measurement are integral components of quantum mechanics, shaping the behavior of quantum systems and challenging our classical intuitions. Superposition allows quantum particles to exist in multiple states simultaneously until measured, providing the foundation for quantum computing and other quantum technologies. Measurement, on the other hand, collapses quantum states into definite values, influencing the outcomes and introducing uncertainty. The interplay between superposition and measurement continues to intrigue physicists, engineers, and philosophers as they navigate the enigmatic landscape of the quantum world.

Chapter 3: Quantum Gates and Circuits

Exploring the intricate realm of quantum computing, one encounters a crucial component known as quantum logic gates. These gates serve as the building blocks of quantum circuits, allowing for the manipulation and processing of quantum information. While quantum logic gates share some similarities with classical logic gates, they also exhibit unique properties and behaviors that arise from the principles of quantum mechanics.

At its core, a quantum logic gate is a quantum-mechanical system that operates on one or more qubits, the quantum counterparts of classical bits. Unlike classical bits, which can only represent 0 or 1, qubits can exist in superpositions of both 0 and 1 states, thanks to the principle of superposition. This property enables quantum logic gates to perform operations that are fundamentally different from classical gates.

One of the most fundamental quantum logic gates is the quantum NOT gate, often denoted as X gate. This gate flips the state of a qubit, transforming $|0\rangle$ into $|1\rangle$ and $|1\rangle$ into $|0\rangle$. While conceptually similar to the classical NOT gate, the quantum NOT gate can operate on qubits in superposition, creating complex interference patterns and enabling quantum algorithms to perform operations on multiple states simultaneously.

Another essential quantum logic gate is the quantum Hadamard gate, represented as H gate. The Hadamard gate plays a pivotal role in creating and manipulating superpositions. When applied to a $|0\rangle$ state, the H gate transforms it into an equal superposition of $|0\rangle$ and $|1\rangle$, represented as $(|0\rangle + |1\rangle) / \sqrt{2}$. Conversely, when applied to

$|1\rangle$, it creates an equal superposition of $|0\rangle$ and $|1\rangle$ with a phase difference, $(|0\rangle - |1\rangle) / \sqrt{2}$. This property is central to quantum algorithms like Grover's algorithm and quantum key distribution.

Quantum logic gates can also perform controlled operations, where the operation is applied to a target qubit based on the state of one or more control qubits. The quantum CNOT (controlled-NOT) gate, for example, flips the state of the target qubit if and only if the control qubit is in the $|1\rangle$ state. This gate serves as a foundational building block for quantum error correction and entanglement generation.

In addition to single-qubit and controlled gates, quantum logic gates include multi-qubit gates that operate on multiple qubits simultaneously. The quantum Toffoli gate, for instance, is a three-qubit gate that performs a controlled-controlled-NOT operation. It flips the target qubit if and only if both control qubits are in the $|1\rangle$ state. Such multi-qubit gates are essential for implementing complex quantum algorithms and quantum error-correcting codes.

Quantum logic gates, like their classical counterparts, can be combined to create quantum circuits. These circuits are composed of a sequence of gates that transform the initial state of the qubits into the desired final state. Quantum circuits are the quantum analogs of classical circuits and serve as the blueprint for quantum algorithms and computations.

One of the most famous quantum algorithms, Shor's algorithm, showcases the power of quantum logic gates. Shor's algorithm can factor large numbers exponentially faster than the best-known classical algorithms, posing a significant threat to classical encryption methods. This algorithm relies on a combination of quantum gates and principles, including quantum superposition and the *quantum Fourier transform, to efficiently factor large numbers into their prime factors.*

Quantum logic gates are not limited to theoretical constructs but have also been implemented in various physical systems. Superconducting qubits, trapped ions, and photonic qubits are among the leading candidates for realizing quantum gates in experimental setups. These implementations require precise control over the quantum states of qubits and the application of gate operations with high fidelity.

The development of fault-tolerant quantum computing relies on the creation of error-correcting codes and the design of fault-tolerant quantum gates. Error-correcting codes protect quantum information from errors and decoherence, ensuring the reliability of quantum computations. Fault-tolerant quantum gates are designed to operate in the presence of errors and noise, making quantum computations robust and scalable.

Quantum logic gates also play a significant role in quantum teleportation and quantum cryptography. Quantum teleportation allows for the transfer of quantum states from one location to another through the use of entangled particles and quantum gates. Quantum cryptography relies on the principles of quantum entanglement and gate operations to ensure secure communication and unbreakable encryption.

In summary, quantum logic gates are the fundamental components of quantum circuits, enabling the manipulation and processing of quantum information. These gates operate on qubits, which can exist in superpositions of states, and they exhibit unique behaviors and capabilities rooted in the principles of quantum mechanics. Quantum logic gates are at the heart of quantum algorithms, quantum error correction, and quantum technologies, with the potential to revolutionize computation and communication in the quantum era. In the realm of quantum computing, building quantum circuits serves as a critical step in harnessing the power of quantum mechanics for solving complex problems. Quantum circuits are the quantum analogs of classical circuits, consisting of quantum logic gates and qubits, and they provide a blueprint for executing quantum algorithms. To understand the process of building quantum circuits, one must delve into the principles of quantum logic gates and their application in quantum computation.

At the core of quantum circuits are quantum logic gates, which are analogous to classical logic gates in traditional digital circuits. These gates perform operations on qubits, the quantum counterparts of classical bits, to manipulate and process quantum information. While quantum logic gates share some similarities with their classical counterparts, they also exhibit unique behaviors that stem from the principles of quantum mechanics.

One of the fundamental quantum logic gates is the quantum NOT gate, also known as the X gate. Similar to the classical NOT gate, the quantum X gate flips the state of a qubit. It transforms $|0\rangle$ into $|1\rangle$ and $|1\rangle$ into $|0\rangle$. However, what sets the quantum X gate apart is its ability to operate on qubits in superposition, allowing for complex interference patterns and simultaneous manipulation of multiple states.

Another essential quantum logic gate is the quantum Hadamard gate, represented as the H gate. The Hadamard gate plays a pivotal role in creating and manipulating superpositions. When applied to a $|0\rangle$ state, the H gate transforms it into an equal superposition of $|0\rangle$ and $|1\rangle$, denoted as $(|0\rangle + |1\rangle) / \sqrt{2}$. Similarly, when applied to $|1\rangle$, it creates an equal superposition with a phase difference, $(|0\rangle - |1\rangle) / \sqrt{2}$. This property is central to quantum algorithms, such as Grover's algorithm and quantum key distribution.

Quantum circuits can also incorporate controlled gates, where the operation is applied to a target qubit based on the state of one or more control qubits. The quantum CNOT (controlled-NOT) gate is a foundational example of such gates. It flips the state of the target qubit if and only if the control qubit is in the |1⟩ state. Controlled gates are vital for creating entanglement and implementing quantum error correction.

In addition to single-qubit and controlled gates, quantum circuits encompass multi-qubit gates that operate on multiple qubits simultaneously. The quantum Toffoli gate, for instance, is a three-qubit gate that performs a controlled-controlled-NOT operation. It flips the target qubit if and only if both control qubits are in the |1⟩ state. Such multi-qubit gates are crucial for executing complex quantum algorithms and quantum error-correcting codes.

Building quantum circuits involves assembling a sequence of quantum gates to achieve a specific computational task. These circuits are akin to the logical flow of instructions in classical computer programs. Quantum algorithms are expressed as sequences of quantum gates, and executing them requires precise control over the quantum states of qubits and the application of gate operations with high fidelity.

Quantum circuit design is not limited to theoretical constructs but has practical implications in the development of quantum technologies. Various physical systems, such as superconducting qubits, trapped ions, and photonic qubits, serve as platforms for realizing quantum gates in experimental setups. Implementing quantum gates in these systems demands meticulous engineering and control techniques to maintain qubit coherence and achieve the desired gate operations.

The development of fault-tolerant quantum computing relies on the creation of error-correcting codes and the design of fault-tolerant quantum gates. Error-correcting codes protect quantum information from errors and decoherence, ensuring the reliability of quantum computations. Fault-tolerant quantum gates are designed to operate in the presence of errors and noise, making quantum computations robust and scalable.

Quantum circuits are the foundation of quantum algorithms, the heart of quantum computing. Shor's algorithm, for instance, showcases the power of quantum circuits by efficiently factoring large numbers, a task that would take classical computers exponentially longer to complete. This algorithm exploits the principles of quantum superposition and phase estimation to revolutionize cryptography and encryption.

Quantum circuits also have applications in quantum simulations, where they mimic the behavior of quantum systems that are difficult to study classically. Quantum chemistry simulations, for example, use quantum circuits to model the behavior of molecules and complex chemical reactions, offering insights into drug discovery and materials science.

In summary, building quantum circuits is a fundamental step in harnessing the potential of quantum computing. These circuits consist of quantum logic gates that manipulate qubits, leveraging the principles of quantum mechanics to perform complex computations. Quantum circuits are the foundation of quantum algorithms, enabling advancements in cryptography, quantum simulations, and various other fields. As we continue to explore the capabilities of quantum circuits, they remain at the forefront of cutting-edge research and technological innovation.

Chapter 4: Entanglement: The Quantum Connection

In the intriguing landscape of quantum physics, the EPR paradox and Bell's theorem stand as two pivotal concepts that challenge our understanding of the quantum world. The EPR paradox, named after its creators Einstein, Podolsky, and Rosen, emerged in 1935 as a thought experiment designed to question the completeness of quantum mechanics. It raised profound questions about the nature of reality and entanglement, setting the stage for Bell's theorem, proposed by physicist John Bell in 1964, which provided a way to experimentally test the predictions of quantum mechanics and address the issues raised by the EPR paradox.

The EPR paradox begins with the notion of entanglement, a fundamental phenomenon in quantum mechanics. Entanglement occurs when two or more particles become correlated in such a way that the measurement of one particle instantly influences the state of the other, even if they are separated by vast distances. This eerie interconnectedness defies classical notions of locality and challenges our intuitive understanding of how information can travel.

Einstein, Podolsky, and Rosen proposed a scenario in which two particles, say electrons, are created in such a way that their properties are correlated. According to quantum mechanics, the properties of one electron, such as its spin, are not determined until they are measured. However, the EPR argument suggested that if the properties of one electron were measured, the properties of the other electron would be instantly known, even if it were far away. This implied a form of "spooky action at a distance," as Einstein put it, that seemed to violate the principle of locality.

The paradox raised by EPR was whether quantum mechanics provided a complete description of physical reality, as it seemed to allow for non-local influences between particles. Einstein was uncomfortable with the implications of quantum mechanics and believed that hidden variables, yet to be discovered, could provide a more deterministic and local explanation for quantum phenomena.

Bell's theorem addressed the EPR paradox by introducing a mathematical framework to test the predictions of quantum mechanics against the principles of local realism. Local realism posits that physical systems have pre-existing properties that determine their behavior, and influences cannot propagate faster than the speed of light, adhering to classical notions of causality and locality.

Bell's theorem introduced a set of inequalities, known as Bell inequalities, which could be tested experimentally. Violation of these inequalities would suggest that the predictions of quantum mechanics could not be explained by local realism, indicating the presence of non-local correlations or entanglement.

Experiments conducted to test Bell's inequalities consistently showed violations, providing empirical evidence that quantum mechanics did indeed exhibit non-local correlations and that local realism was not a viable explanation for the behavior of entangled particles. This result was a remarkable confirmation of the predictions of quantum mechanics and a departure from classical physics.

Bell's theorem showed that the EPR paradox could not be resolved by hidden variables or local realism, as Einstein had hoped. Instead, it underscored the non-classical nature of quantum mechanics, where entanglement allows for correlations that transcend classical boundaries of space and time. This concept challenged our intuitive understanding of *causality and locality, raising questions about the fundamental nature of reality.*

The experimental confirmation of Bell's theorem had profound implications for the field of quantum physics. It highlighted the existence of entanglement as a fundamental and non-negotiable aspect of quantum mechanics. It also indicated that the behavior of quantum systems could not be explained by classical determinism or hidden variables, reinforcing the probabilistic and non-local nature of quantum phenomena.

The concept of entanglement, once regarded as a theoretical curiosity, has since become a central focus of quantum research. It has led to the development of quantum technologies such as quantum cryptography and quantum computing, where entangled particles can be harnessed for secure communication and exponential computational speedup.

Moreover, Bell's theorem has stimulated ongoing philosophical debates about the nature of reality and the role of measurement in quantum mechanics. It challenges our classical intuitions and invites us to rethink our assumptions about causality, determinism, and the fundamental structure of the universe.

In summary, the EPR paradox and Bell's theorem represent two significant milestones in the history of quantum physics. The EPR paradox raised questions about the completeness of quantum mechanics and the nature of entanglement, while Bell's theorem provided a framework to experimentally test the principles of local realism and confirmed the non-classical nature of quantum correlations. These concepts continue to shape our understanding of the quantum world and inspire further exploration into the mysteries of quantum mechanics.

Exploring the practical implications of quantum entanglement, one enters a realm of astonishing possibilities and technological advancements. Quantum entanglement, a phenomenon first described by Einstein, Podolsky, and Rosen in their famous EPR paper, involves the correlation of quantum states between particles, regardless of the distance that separates them. While it initially sparked debates and perplexed physicists, quantum entanglement has evolved from a theoretical concept into a tangible resource with applications that span quantum communication, cryptography, and emerging technologies.

Quantum entanglement challenges our classical intuition by enabling particles to become interconnected in ways that transcend classical boundaries. When two particles are entangled, measuring the state of one instantly determines the state of the other, even if they are light-years apart. This instantaneous connection, famously referred to by Einstein as "spooky action at a distance," intrigued scientists and led to a deeper exploration of the quantum world.

One of the practical applications of quantum entanglement is in quantum communication. Quantum communication harnesses the properties of entangled particles to achieve a level of security that classical communication methods cannot provide. Quantum key distribution (QKD) is a prime example of this. In QKD, two parties share entangled particles and use them to create a secret cryptographic key. Any attempt to intercept the key will inevitably disturb the entangled particles, revealing the eavesdropper's presence and ensuring the security of the communication.

Quantum entanglement has already found its way into the world of secure communications. Commercial QKD systems are available, and quantum-secure communication networks are being developed to protect sensitive information from eavesdropping, ensuring the confidentiality of data transmitted over long distances.

Another intriguing application of quantum entanglement is in quantum teleportation. While it doesn't involve physically moving objects from one place to another, quantum teleportation allows for the transfer of quantum states between entangled particles. This process relies on entangled particles to transmit the quantum information required to recreate a quantum state at a remote location. Quantum teleportation holds potential in quantum computing, where quantum states need to be transferred between qubits in a fault-tolerant quantum computer.

Quantum entanglement also plays a pivotal role in quantum cryptography. Beyond quantum key distribution, quantum entanglement-based protocols, such as the BBM92 protocol, enable secure multi-party communication, ensuring the privacy of group conversations in quantum networks. These cryptographic advancements have the potential to reshape the landscape of secure communications in a quantum world.

The concept of quantum entanglement has even extended into the realm of quantum computing, where it forms the foundation of quantum algorithms. Algorithms like Shor's algorithm and Grover's algorithm leverage the power of entanglement to solve problems exponentially faster than classical computers. Shor's algorithm, for instance, efficiently factors large numbers, posing a significant threat to classical encryption methods. Grover's algorithm, on the other hand, accelerates database search and optimization tasks, promising transformative applications in data analysis and cryptography.

In practice, the creation and maintenance of entangled particles for quantum computing pose significant challenges. Researchers are exploring various physical systems, including superconducting qubits, trapped ions, and photonic qubits, to generate and manipulate entanglement with high fidelity. These efforts aim to harness entanglement's potential to revolutionize computation and problem-solving.

Quantum entanglement has also found applications in the emerging field of quantum-enhanced sensors. Entangled particles can be used to create precision instruments that surpass the sensitivity of classical sensors. For instance, quantum entanglement can enhance the precision of atomic clocks, gyroscopes, and magnetometers, enabling advancements in navigation, timekeeping, and scientific measurements.

Furthermore, quantum entanglement has been investigated in the context of quantum simulations. By entangling particles in a controlled manner, researchers can simulate the behavior of quantum systems that are otherwise challenging to study classically. Quantum chemistry simulations, for example, use entangled qubits to model the behavior of molecules and chemical reactions, offering insights into drug discovery, materials science, and fundamental chemistry.

The practical applications of quantum entanglement extend into the realm of quantum metrology. Quantum metrology leverages entanglement to improve the precision of measurements, making it possible to detect minute changes in physical quantities such as temperature, pressure, and magnetic fields. These advancements have implications for a wide range of industries, from healthcare to environmental monitoring.

Moreover, quantum entanglement has been explored in the context of quantum imaging and quantum-enhanced imaging techniques. Quantum-enhanced sensors and imagers promise to revolutionize fields such as medical imaging, remote sensing, and security screening, where high-resolution and sensitivity are critical.

In the field of quantum networks, entanglement-based quantum repeaters have been proposed to extend the range of secure quantum communication. Quantum repeaters use entanglement to transmit quantum information over long distances, overcoming the limitations imposed by the attenuation of optical signals in optical fibers. This technology has the potential to create global-scale quantum communication networks, enabling secure communication and quantum-enhanced capabilities on a global scale.

In summary, quantum entanglement has transcended its theoretical origins to become a practical resource with a wide range of applications. From secure quantum communication and cryptography to quantum computing, sensors, simulations, and imaging, entanglement continues to drive innovations that have the potential to transform technology, science, and industry. The profound interconnectedness of entangled particles has become a cornerstone of the quantum world, paving the way for a new era of quantum technologies and capabilities.

Chapter 5: Quantum Algorithms and their Applications

Shor's algorithm, a groundbreaking quantum algorithm, has the power to revolutionize cryptography and computational mathematics by solving a problem that has remained formidable for classical computers—integer factorization. The ability to efficiently factor large numbers is of paramount importance in cryptography, as it underpins the security of widely-used encryption schemes. The security of many cryptographic systems, including the RSA algorithm, relies on the fact that factoring large semiprime numbers is a time-consuming task for classical computers. However, Shor's algorithm threatens to render these classical encryption methods vulnerable.

The problem of integer factorization involves decomposing a composite number into its prime factors. For example, given the number 21, the prime factorization would be 3 × 7. While this may seem straightforward for small numbers, it becomes exceedingly challenging as the number to be factored grows larger, particularly when dealing with numbers that have hundreds or thousands of digits.

Classical algorithms for integer factorization, such as the trial division method or the general number field sieve, become exponentially slower as the size of the number increases. This exponential growth in complexity makes it practically infeasible to factor large numbers in a reasonable amount of time using classical computers.

Enter Shor's algorithm, conceived by mathematician Peter Shor in 1994. Shor's algorithm leverages the principles of quantum mechanics to factor large numbers exponentially faster than the best-known classical algorithms. Its groundbreaking nature lies in its ability to exploit the quantum properties of superposition and entanglement.

The key to Shor's algorithm's speed lies in its quantum Fourier transform. By employing a quantum version of the classical Fourier transform, Shor's algorithm efficiently finds the period of a function, which is a crucial step in factoring large numbers. This period-finding capability is what sets Shor's algorithm apart from classical factorization methods.

Shor's algorithm begins by representing the number to be factored as a quantum state. This state is then manipulated using quantum gates, including modular exponentiation, which is a quantum operation that exploits the periodicity of the function related to factorization. The quantum Fourier transform is applied to extract the period of the function, revealing the factors of the original number.

The most remarkable aspect of Shor's algorithm is its efficiency. While the best classical algorithms require exponential time to factor large numbers, Shor's algorithm operates in polynomial time. This means that as the size of the number to be factored increases, the time required to factor it using Shor's algorithm increases much more slowly than with classical methods.

Shor's algorithm is particularly threatening to the security of widely-used encryption schemes that rely on the difficulty of integer factorization. The RSA algorithm, for instance, is based on the assumption that factoring the product of two large prime numbers is a computationally infeasible task. However, with the advent of large-scale quantum computers capable of running Shor's algorithm, the security of RSA and similar encryption methods would be compromised.

The potential implications of Shor's algorithm for cryptography have spurred research into quantum-resistant encryption schemes. Post-quantum cryptography seeks to develop encryption algorithms that would remain secure even in the presence of quantum computers with Shor's algorithm capabilities. These post-quantum encryption methods rely on mathematical problems that are believed to be difficult for both classical and quantum computers to solve.

Quantum-resistant encryption is essential to protect sensitive information and communications from being decrypted by future quantum computers. The transition to quantum-resistant cryptography is a critical aspect of ensuring the security of digital communication and data storage in a quantum-powered world.

Beyond its implications for cryptography, Shor's algorithm has implications for computational mathematics and number theory. It provides a new perspective on the nature of factoring large numbers and the relationship between classical and quantum computation. Shor's algorithm has sparked interest in exploring the potential of quantum computers for solving other mathematical problems more efficiently than classical computers.

However, it's worth noting that, as of my last knowledge update in 2022, the practical implementation of large-scale quantum computers capable of running Shor's algorithm remains a significant challenge. Building and maintaining stable, error-corrected quantum computers with the required number of qubits is an ongoing research effort.

In summary, Shor's algorithm represents a major milestone in the field of quantum computing. It has the potential to disrupt classical encryption methods that rely on the difficulty of integer factorization, prompting the development of quantum-resistant encryption schemes. Shor's algorithm's efficient approach to factoring large numbers is a testament to the power of quantum computation, offering a glimpse into the transformative capabilities of quantum computing in various fields beyond cryptography. As the development of quantum technology progresses, the implications of Shor's algorithm continue to shape the future of digital security and computational mathematics.

Grover's algorithm, a quantum algorithm devised by Lov Grover in 1996, represents a milestone in quantum computing for its exceptional ability to perform unstructured search problems exponentially faster than classical algorithms. This algorithm addresses a fundamental problem known as the unstructured search problem, which involves finding a specific item within an unsorted database or an unordered list. Grover's algorithm achieves remarkable efficiency by utilizing the principles of quantum superposition and interference.

The unstructured search problem, often exemplified as searching for a marked item in an unsorted list, serves as a common task in computer science and information retrieval. In classical computing, solving this problem typically requires examining each item in the list one by one, making it a time-consuming process with a complexity proportional to the number of items in the list. For a list of N items, classical algorithms typically require $O(N)$ time to find the desired item.

Grover's algorithm, in stark contrast, achieves quadratic speedup over classical algorithms. It can find the marked item in a list of N items in roughly $\sqrt{N}$ steps, significantly reducing the time required for the search. This quadratic speedup holds immense significance for a wide range of applications, particularly in areas where searching for specific information quickly is crucial, such as database searches, cryptography, and optimization problems.

The key to Grover's algorithm's efficiency lies in its clever use of quantum mechanics. The algorithm starts by creating a quantum superposition of all possible states representing items in the unsorted list. In other words, it processes all potential solutions simultaneously. This quantum superposition allows Grover's algorithm to explore multiple states in parallel, a capability unavailable to classical algorithms.

Grover's algorithm then employs a series of quantum operations, including oracle queries and quantum gates, to amplify the amplitude of the state corresponding to the marked item while decreasing the amplitudes of the other states. The oracle query is a crucial component of Grover's algorithm, as it serves to identify the marked item and create constructive interference that enhances the probability of finding it.

As the algorithm iterates through these quantum operations, it systematically increases the probability of measuring the state representing the marked item. After approximately $\sqrt{N}$ iterations, Grover's algorithm reaches a point of high amplitude for the marked item, ensuring a high probability of successful measurement.

One remarkable feature of Grover's algorithm is its universality. It can be applied to a wide range of search problems without prior knowledge of the item's location or the structure of the list. Grover's algorithm efficiently solves unstructured search problems in cases where classical algorithms would require exhaustive searches.

While Grover's algorithm's primary application is quantum search, it has also been extended to solve other computational tasks, such as database search, satisfiability problems, and graph-related problems. These adaptations showcase the algorithm's versatility and potential to address various real-world challenges efficiently.

One notable application of Grover's algorithm is in cryptography. It has implications for breaking symmetric-key encryption schemes, where the key space can be searched to find the correct decryption key. Grover's algorithm's quadratic speedup poses a significant threat to classical encryption methods, prompting the need for larger key sizes to maintain security in a post-quantum world.

Conversely, Grover's algorithm can be used for quantum-enhanced cryptography. It allows for the design of quantum-resistant encryption methods by utilizing the same principles of quantum search to create encryption schemes that are secure against quantum attacks.

The efficiency of Grover's algorithm also holds promise in optimization problems. It can be adapted to find the optimal solutions to combinatorial optimization problems more quickly than classical algorithms. This capability has implications for a wide range of fields, from logistics and supply chain management to artificial intelligence and machine learning.

In practice, implementing Grover's algorithm on a quantum computer requires precise control over the quantum states and quantum gates. Quantum computers, which operate on quantum bits or qubits, are still in the early stages of development, and building large-scale quantum computers capable of running Grover's algorithm efficiently remains a significant challenge.

However, as quantum technology advances, Grover's algorithm is expected to play a crucial role in quantum computing applications. It exemplifies the power of quantum computation to provide exponential speedup for specific tasks, offering a glimpse into the transformative capabilities of quantum computers.

In summary, Grover's algorithm represents a remarkable achievement in the field of quantum computing, showcasing the *potential for exponential speedup in solving unstructured search problems. Its efficiency has applications in cryptography, optimization, and various computational tasks, making it a key algorithm in the emerging field of quantum computing. As quantum technology continues to evolve, Grover's algorithm holds promise for addressing complex problems efficiently and revolutionizing fields that rely on fast and efficient searching and optimization.*

Chapter 6: Building Your First Quantum Program

Quantum programming languages represent a bridge between the abstract principles of quantum mechanics and the practical implementation of quantum algorithms. These specialized programming languages enable researchers and developers to express quantum algorithms in a way that can be executed on quantum computers. Unlike classical programming languages that operate on bits, quantum programming languages deal with qubits, the fundamental units of quantum information.

At the core of quantum programming languages is the concept of quantum superposition, where qubits can exist in multiple states simultaneously. This property allows quantum algorithms to perform certain calculations exponentially faster than classical algorithms. To harness this potential, quantum programming languages provide a framework for expressing quantum operations, transformations, and measurements.

One of the most widely-used quantum programming languages is Qiskit, developed by IBM. Qiskit is an open-source framework that allows users to create, manipulate, and execute quantum circuits. It provides a high-level interface for defining quantum algorithms, making it accessible to both beginners and experienced quantum researchers.

Another prominent quantum programming language is Quipper, developed at Microsoft Research and the University of Oxford. Quipper is a functional programming language designed for expressing quantum algorithms. It allows for a high degree of abstraction and can be compiled to various quantum platforms.

Microsoft's Q# is yet another quantum programming language that is part of the Quantum Development Kit. Q# is designed to work seamlessly with classical programming languages like C#

and allows for the creation of quantum algorithms, simulations, and quantum libraries.

Python-based languages are also gaining popularity in the quantum programming domain. PyQuil, for example, is a quantum programming framework developed by Rigetti Computing. It combines the power of Python with the ability to define quantum operations and execute them on quantum processors.

Quantum programming languages often include libraries and tools for quantum simulation. These simulators allow developers to test and debug quantum algorithms on classical computers before running them on actual quantum hardware. This simulation capability is essential for the development and validation of quantum algorithms.

Quantum programming languages also offer features for error correction and fault tolerance. Quantum computers are susceptible to errors due to factors such as decoherence and noise. Error-correcting codes and fault-tolerant techniques are essential for ensuring the reliability of quantum computations. Quantum programming languages provide constructs for encoding and decoding quantum information to mitigate errors.

Programming quantum computers requires a deep understanding of quantum mechanics, as well as knowledge of the specific language and framework being used. Quantum algorithms are expressed as quantum circuits, which are composed of quantum gates that manipulate qubits. These gates can include operations like quantum NOT gates, quantum Hadamard gates, and custom-defined gates.

The design of quantum algorithms often involves leveraging quantum properties such as superposition and entanglement to perform calculations efficiently. Quantum algorithms like Grover's algorithm and Shor's algorithm showcase the power of quantum programming in solving problems exponentially faster than classical counterparts.

Quantum programming languages are not only used for theoretical research but also for practical applications. Quantum machine learning, for example, is an emerging field that combines quantum computing and classical machine learning techniques. Quantum programming languages play a crucial role in implementing quantum machine learning algorithms.

Quantum cryptography is another area where quantum programming languages are applied. Quantum key distribution protocols, like BB84, rely on the secure exchange of quantum information. Quantum programming languages are used to implement these protocols and ensure the security of quantum communication.

The development of quantum hardware is closely tied to the evolution of quantum programming languages. As quantum processors become more accessible, quantum programming languages need to adapt to support various quantum architectures and provide tools for optimizing quantum circuits.

Quantum programming languages are not limited to specific industries or research domains. They have applications in fields as diverse as materials science, finance, optimization, and quantum chemistry. Quantum simulations, for instance, are used to study the behavior of molecules and materials, leading to advancements in drug discovery and materials engineering.

In recent years, cloud-based quantum computing platforms have emerged, allowing users to access quantum processors remotely. These platforms often provide integrated development environments (IDEs) with quantum programming capabilities. Users can write quantum code, submit it to the cloud-based quantum computer, and receive the results, all through the platform's interface.

Quantum programming languages are continuously evolving as the field of quantum computing progresses. New languages, libraries, and tools are being developed to simplify quantum programming and make it more accessible to a broader *audience. Quantum software development kits (SDKs) and quantum compilers are among the innovations that streamline the process of translating quantum algorithms into executable code.*

As quantum computing technology matures, quantum programming languages will become increasingly integral to harnessing the potential of quantum machines. Their role in shaping the future of quantum computing cannot be overstated, as they provide the means to express complex quantum algorithms, test them, and ultimately deploy them for practical applications in a wide range of fields. The development and adoption of quantum programming languages represent a critical step toward realizing the full potential of quantum computing in our increasingly digital world.

Creating simple quantum programs marks the first step on the journey to harnessing the power of quantum computing. While the field of quantum computing may seem daunting at first, starting with basic quantum programs provides an accessible entry point. Quantum programs are designed to leverage the unique properties of quantum bits or qubits, such as superposition and entanglement, to perform computations in ways that classical computers cannot.

Before diving into quantum programming, it's essential to understand the fundamental building blocks of quantum computation. At the heart of quantum computing are qubits, which are the quantum analogs of classical bits. While classical bits can be either 0 or 1, qubits can exist in a superposition of both states simultaneously.

This property of superposition allows quantum programs to explore multiple computational paths in parallel. However, measuring a qubit in superposition collapses it into one of the classical states, 0 or 1, with a probability determined by the amplitudes of the superposition.

Entanglement is another crucial quantum concept. When two qubits become entangled, the measurement of one qubit instantaneously determines the state of the other, regardless of the distance separating them. This non-local correlation is a hallmark of quantum mechanics and serves as a powerful resource in quantum programming.

To start creating simple quantum programs, you'll need access to a quantum programming environment or framework. There are several quantum development platforms available, such as IBM's Qiskit, Microsoft's Quantum Development Kit, Google's Cirq, and Rigetti's Forest. These platforms provide the tools and libraries necessary to write, simulate, and run quantum programs.

Let's take a look at a simple quantum program in Qiskit, one of the most popular quantum programming frameworks. Suppose we want to create a quantum circuit that puts a qubit in a superposition of states $|0\rangle$ and $|1\rangle$.

pythonCopy code

```
from qiskit import QuantumCircuit, Aer, transpile, assemble # Create a quantum circuit with one qubit qc = QuantumCircuit(1) # Apply a Hadamard gate to put the qubit in superposition qc.h(0) # Simulate the quantum circuit simulator = Aer.get_backend('aer_simulator') compiled_circuit = transpile(qc, simulator) job = assemble(compiled_circuit) result = simulator.run(job).result() # Measure the qubit counts = result.get_counts() print(counts)
```

In this example, we import the necessary modules from Qiskit and create a quantum circuit with one qubit. We then apply a Hadamard gate (represented by qc.h(0)) to put the qubit into a superposition of $|0\rangle$ and $|1\rangle$ states. The next steps involve simulating the quantum circuit and measuring the qubit.

Simulating quantum programs is a critical step in quantum development, as it allows you to understand and verify the behavior of your quantum circuits before running them on real quantum hardware. Quantum simulators, such as the one used *in this example, provide a classical approximation of quantum behavior.*

When you run this simple quantum program, you'll notice that the measurement results may vary between |0⟩ and |1⟩ with approximately equal probabilities, demonstrating the effect of the Hadamard gate's superposition operation.

Creating more complex quantum programs involves designing quantum circuits with multiple qubits and applying various quantum gates to manipulate their states. Quantum gates are analogous to classical logical gates but operate on qubits in a quantum superposition.

For instance, the CNOT gate, or controlled-X gate, is a common quantum gate that introduces entanglement between qubits. You can use it to create entangled pairs of qubits, a phenomenon known as Bell pairs. Here's an example of how you can create a Bell pair in Qiskit:

pythonCopy code

from qiskit import QuantumCircuit, Aer, transpile, assemble # Create a quantum circuit with two qubits qc = QuantumCircuit(2) # Apply a Hadamard gate to the first qubit qc.h(0) # Apply a CNOT gate to create a Bell pair qc.cx(0, 1) # Simulate the quantum circuit simulator = Aer.get_backend('aer_simulator') compiled_circuit = transpile(qc, simulator) job = assemble(compiled_circuit) result = simulator.run(job).result() # Measure both qubits counts = result.get_counts() print(counts)

In this example, we create a quantum circuit with two qubits, apply a Hadamard gate to the first qubit to put it in superposition, and then apply a CNOT gate to create entanglement between the two qubits. The resulting measurement outcomes will demonstrate entanglement between the qubits.

As you delve deeper into quantum programming, you'll encounter a variety of quantum gates, each with its specific role and effect on qubits. Some gates introduce phase shifts, while others perform conditional operations based on the state of one or more qubits.

Quantum programming languages also provide features for error correction and fault tolerance, which are essential for reliable quantum computations. Error correction codes, such as the surface code, help mitigate the effects of noise and decoherence that can occur in quantum hardware.

Quantum development platforms often offer quantum cloud services that allow you to run your quantum programs on actual quantum processors. Quantum cloud services provide access to remote quantum computers, enabling you to execute your quantum algorithms on the latest quantum hardware.

While creating simple quantum programs is a fascinating endeavor, it's important to note that quantum programming is still an evolving field. Quantum hardware is continually advancing, and quantum programming languages and frameworks are adapting to support new capabilities and technologies.

In summary, creating simple quantum programs is an exciting journey into the world of quantum computing. It allows you to explore the unique properties of qubits, such as superposition and entanglement, and gain hands-on experience with quantum programming languages and frameworks. As you continue to develop your quantum programming skills, you'll have the opportunity to tackle increasingly complex quantum algorithms and contribute to the advancement of quantum computing technologies.

Chapter 7: Quantum Error Correction

Understanding the various types of quantum errors is fundamental in the pursuit of robust and fault-tolerant quantum computation. Quantum computers hold immense potential, but they are susceptible to errors due to their sensitivity to external influences and the delicate nature of quantum states. These errors can undermine the reliability of quantum algorithms, making it crucial to develop strategies for detecting, correcting, and mitigating them.

One of the primary types of quantum errors is quantum decoherence. Decoherence occurs when a quantum system loses its coherence, which is the property that allows quantum states, like superpositions, to persist over time. Decoherence is typically caused by interactions with the environment, including factors such as temperature, electromagnetic radiation, and even stray magnetic fields. As qubits lose their coherence, the information they carry becomes unreliable, hindering the execution of quantum algorithms.

Another significant source of quantum errors is quantum gate errors. Quantum gates are the fundamental operations that manipulate qubits to perform computations. These gates are not perfect, and errors can occur during their application. For example, a quantum gate might introduce a phase error, flipping a qubit's state or causing it to deviate from the desired computation path. Minimizing gate errors is a critical aspect of quantum error correction.

Furthermore, qubit errors are prevalent in quantum systems. Qubit errors can stem from various sources, including imperfections in the hardware used to implement qubits. These hardware imperfections may lead to variations in qubit energy levels or couplings between qubits, resulting in errors during quantum operations. Managing and mitigating qubit errors is essential for reliable quantum computation.

Gate duration errors represent another class of quantum errors. In quantum computing, gate operations are not instantaneous; they take time to execute. During this execution time, the quantum system can be exposed to environmental factors that may lead to errors. Managing gate duration errors involves optimizing gate times and minimizing the impact of external influences during gate operations.

Quantum measurement errors also play a role in quantum computing. When a quantum system is measured, the measurement outcome can be affected by various factors, such as measurement device imperfections or the entanglement of the measured qubit with its environment. Correcting measurement errors is essential for obtaining accurate results from quantum computations.

Another quantum error category is state preparation errors. Preparing qubits in specific initial states is a crucial step in many quantum algorithms. Errors in state preparation can occur due to imperfect control over qubit initializations, leading to deviations from the desired quantum states. Developing techniques to enhance state preparation accuracy is crucial for reliable quantum computing.

In addition to the aforementioned errors, quantum systems can also be affected by readout errors. Readout errors occur when measuring the final state of a quantum computation. These errors can be caused by imperfections in the measurement apparatus or the presence of unwanted environmental factors. Addressing readout errors is vital for obtaining meaningful and accurate results from quantum computations.

Quantum error correction (QEC) is a field of research dedicated to mitigating quantum errors and ensuring the reliability of quantum computations. QEC employs specialized codes, such as the surface code and the Steane code, to detect and correct errors that occur during quantum operations. These codes introduce redundancy into the quantum information, allowing for the identification and rectification of errors.

Fault-tolerant quantum computing is an ambitious goal that aims to achieve reliable quantum computation even in the presence of errors. This approach involves designing quantum algorithms and hardware architectures that can withstand a certain level of errors without compromising the overall computation. Fault-tolerant quantum error correction codes, combined with fault-tolerant quantum gates and algorithms, are central to this endeavor.

Another significant development in addressing quantum errors is quantum error mitigation. Quantum error mitigation techniques aim to reduce the impact of errors on quantum computations without relying solely on error correction codes. These techniques include error extrapolation, error mitigation circuits, and variational algorithms designed to suppress errors and improve the accuracy of quantum results.

Quantum error monitoring is an essential aspect of quantum error management. Monitoring involves continuously assessing the state of a quantum system during computation to detect and diagnose errors as they occur. Real-time error monitoring enables quantum computers to adapt and make corrections during the execution of quantum algorithms, enhancing the reliability of results.

Quantum hardware improvements also play a crucial role in error mitigation. Advances in qubit technologies, quantum gate fidelities, and error-protected qubit designs contribute to reducing the overall error rates in quantum computations. Ongoing research in quantum hardware aims to create more robust and error-resistant quantum devices.

In summary, quantum errors are a significant challenge in the development of practical quantum computers. Various types of errors, including decoherence, gate errors, qubit errors, gate duration errors, measurement errors, state preparation errors, and readout errors, can affect the reliability of quantum computations. Mitigating and managing these errors is a central focus of quantum error correction, fault-tolerant computing, error mitigation techniques, and continuous monitoring of quantum systems. As quantum technology advances, addressing quantum errors will be critical for unlocking the full potential of quantum computing in various fields, from cryptography to materials science and optimization.

Quantum error correction codes represent a pivotal advancement in the field of quantum computing, providing a pathway to overcome the inherent fragility of quantum information. In the quest to build practical and reliable quantum computers, it's crucial to address the persistent challenge of quantum errors, which can arise from environmental factors, hardware imperfections, and the delicate nature of quantum states. Quantum error correction codes offer a systematic approach to detecting and correcting errors, ensuring the integrity of quantum computations.

At the core of quantum error correction is the idea of encoding quantum information redundantly, allowing for the identification and rectification of errors that may occur during quantum operations. These codes serve as a protective shield around quantum information, safeguarding it from the effects of decoherence and other error sources. By employing quantum error correction codes, researchers and engineers aim to create fault-tolerant quantum computers that can execute complex algorithms with high reliability.

One of the pioneering quantum error correction codes is the quantum repetition code. The repetition code involves encoding a qubit multiple times to form an error-protected logical qubit. For instance, a single qubit can be encoded three times to create a logical qubit that can detect and correct errors. If an error occurs on one of the qubits, the redundant encoding allows the error to be identified and corrected, preserving the integrity of the quantum information.

The surface code, a more advanced quantum error correction code, has gained significant attention for its fault-tolerant properties. The surface code arranges qubits on a two-dimensional lattice, with each qubit interacting with its neighboring qubits. This lattice structure allows for the detection and correction of errors through a process called syndrome measurement. By measuring the parity of qubits in local neighborhoods, the surface code can pinpoint errors and facilitate their correction.

Another notable quantum error correction code is the Steane code, which utilizes seven qubits to protect a single logical qubit. The Steane code corrects errors through a combination of stabilizer measurements and quantum gates, providing a reliable means of error correction. It serves as an essential building block for fault-tolerant quantum computation.

Quantum error correction also involves the use of stabilizer codes, which encompass a broad class of codes capable of detecting and correcting errors. Stabilizer codes are defined by a set of generators, or stabilizers, that commute with the quantum gates used in the code. These stabilizers enable the detection of errors by measuring their eigenvalues, allowing *for the identification of errors that occurred during computation.*

The development of quantum error correction codes extends to the realm of fault-tolerant quantum computing, where quantum algorithms can be executed reliably even in the presence of errors. Fault tolerance relies on the use of concatenated codes, which involve nesting multiple layers of error correction codes. Concatenated codes provide an extra layer of protection against errors, making it possible to achieve fault-tolerant quantum computation.

Quantum error correction is not limited to correcting bit-flip errors but also includes the mitigation of phase-flip errors and combinations of both. The use of quantum gates, such as the CNOT gate and the T gate, is instrumental in implementing quantum error correction codes. These gates allow for the manipulation of qubits to correct errors and maintain the coherence of quantum information.

Despite the promise of quantum error correction, it is essential to acknowledge the challenges associated with its practical implementation. Quantum error correction typically requires a significant overhead in terms of the number of physical qubits needed to protect a single logical qubit. This overhead can be substantial, making it necessary to develop quantum hardware with a sufficient number of qubits to support error correction.

Quantum error correction also demands precise control over qubits and their interactions, as well as the ability to measure syndromes accurately. This level of control and measurement fidelity is a critical requirement for effective error correction in quantum systems.

Furthermore, quantum error correction codes necessitate advanced error models and algorithms for efficient decoding and correction. The development of error-mitigation techniques and strategies to minimize the effects of errors on quantum computations is an ongoing area of research.

Quantum error correction is an interdisciplinary field that intersects quantum information theory, quantum algorithms, quantum hardware, and quantum software development. It relies on a combination of theoretical foundations and practical implementations to achieve its objectives.

In summary, quantum error correction codes are a cornerstone of quantum computing, addressing the challenge of quantum errors that can jeopardize the reliability of quantum computations. These codes, including the repetition code, surface code, Steane code, and stabilizer codes, provide a framework for detecting and correcting errors in quantum systems. The pursuit of fault-tolerant quantum computing, enabled by concatenated codes, promises to unlock the full potential of quantum computers in various fields, from cryptography and materials science to optimization and artificial intelligence. Quantum error correction continues to be a central focus of research and development in the quest for practical and robust quantum computation.

Chapter 8: Quantum Hardware: From Qubits to Quantum Computers

Exploring quantum processor architectures unveils the intricate designs that underpin the operation of quantum computers, devices poised to revolutionize computation as we know it. At the heart of quantum computing lies the quantum processor, responsible for executing quantum algorithms by manipulating quantum bits or qubits. Quantum processor architectures vary widely, with different technologies and approaches employed to harness the unique properties of quantum mechanics.

One of the most common quantum processor architectures is the superconducting qubit-based system. Superconducting qubits are typically fabricated from superconducting materials that exhibit zero electrical resistance when cooled to extremely low temperatures. This architecture leverages the properties of Josephson junctions, which are key components in superconducting qubit circuits.

Superconducting qubits are highly coherent and can be manipulated with microwave pulses. They are connected through superconducting wires to form quantum circuits. Quantum gates, such as CNOT gates and single-qubit gates, are applied to the qubits to perform computations. Prominent companies like IBM, Google, and Rigetti employ superconducting qubits in their quantum processors, which are made available through cloud-based quantum computing platforms.

Trapped-ion quantum processors represent another prominent architecture. In this approach, qubits are encoded in the internal energy levels of trapped ions, typically ytterbium or calcium ions. Laser beams are used to manipulate the ions' quantum states and perform quantum operations. Trapped-ion systems offer long coherence times and high-fidelity quantum gates, making them attractive for quantum computing applications.

Photonic quantum processors are a distinct class of quantum architecture that leverages photons as qubits. Photonic qubits are typically generated and manipulated using optical components, such as beam splitters and phase shifters. Photonic processors are well-suited for tasks like quantum communication and quantum cryptography due to the ease with which photons can be transmitted over long distances.

Topological quantum processors are a burgeoning area of research. These processors rely on topological qubits, which are anyons that exhibit non-Abelian statistics. Topological qubits are theoretically highly robust against errors, making them a promising avenue for building fault-tolerant quantum computers. However, realizing topological qubits and processors remains a significant challenge, and this architecture is in the early stages of development.

Trilinear array architectures are another approach to quantum processors. In a trilinear array, qubits are arranged in a three-dimensional lattice, and quantum gates are performed by coupling neighboring qubits through controlled operations. This architecture offers scalability and fault-tolerant properties when properly implemented. Research is ongoing to optimize trilinear array designs for practical quantum computing.

One common theme across quantum processor architectures is the need for quantum error correction. Quantum processors are susceptible to errors due to factors like decoherence, gate imperfections, and external noise. Quantum error correction codes, such as the surface code and the Steane code, are crucial for detecting and correcting errors, ultimately enabling fault-tolerant quantum computation.

Quantum processor designs often incorporate error mitigation techniques to improve the reliability of quantum computations. These techniques involve characterizing and mitigating errors during quantum algorithm execution. Error mitigation is particularly important when accessing quantum processors with limited qubit counts and gate fidelities.

Quantum processor architectures are evolving rapidly, driven by advances in qubit technologies and quantum hardware. Hybrid architectures, which combine different types of qubits, are emerging as a promising approach to harnessing the strengths of various qubit technologies while mitigating their weaknesses. Quantum coherence times, gate fidelities, and connectivity between qubits are critical factors in evaluating the performance of quantum processor architectures. Achieving high-quality qubits with long coherence times is essential for executing complex quantum algorithms efficiently.

The quest for quantum advantage, the point at which quantum computers outperform classical computers in specific tasks, drives the ongoing development of quantum processor architectures. Quantum advantage has the potential to revolutionize fields such as cryptography, materials science, optimization, and machine learning.

Quantum algorithms, designed to exploit the inherent parallelism and quantum properties of quantum processors, are a central focus in quantum computing research. Shor's algorithm for factoring large numbers and Grover's algorithm for unstructured search are just a few examples of quantum algorithms that promise exponential speedup over their classical counterparts.

Quantum algorithms often involve constructing quantum circuits, which are sequences of quantum gates applied to qubits. The design and optimization of quantum circuits play a pivotal role in achieving quantum advantage. Quantum software development tools and languages, such as Qiskit, Cirq, and Quipper, facilitate the creation of quantum circuits and algorithms.

Quantum processor architectures also face challenges related to error correction and fault tolerance. Building fault-tolerant quantum computers requires substantial overhead in terms of *qubit count and gate resources. Research efforts are focused on developing efficient fault-tolerant techniques to make quantum error correction practical.*

Quantum processor architectures are intimately tied to quantum hardware development. Quantum devices must be meticulously engineered to maintain qubit coherence, minimize gate errors, and support scalable quantum circuits. Cryogenic cooling, precision control electronics, and advanced qubit packaging are integral components of quantum hardware.

The race to develop quantum processors with increasing qubit counts and higher gate fidelities is a defining feature of the quantum computing landscape. Quantum supremacy, the demonstration of a quantum computer solving a problem faster than classical computers, is a significant milestone on this path.

As quantum technology matures, quantum processors will become more accessible through cloud-based quantum computing platforms, making it possible for researchers and developers worldwide to access and experiment with quantum hardware.

In summary, quantum processor architectures encompass a diverse array of designs and technologies, each with its strengths and challenges. Quantum error correction, quantum algorithms, and quantum hardware development are key facets of advancing quantum processors toward practical applications. Quantum computing holds the promise of transforming industries and solving problems that are currently beyond the reach of classical computers, making the exploration and development of quantum processor architectures an exciting and dynamic field of research and innovation.

Delving into the intricacies of quantum computing, it is imperative to explore the fundamental quantum hardware components that enable the execution of quantum algorithms. Quantum hardware components constitute the physical foundation upon which quantum processors and quantum computers are built, facilitating the manipulation and measurement of quantum bits or qubits. These components encompass a wide range of technologies and techniques, each playing a unique role in harnessing the power of quantum mechanics for computation.

At the heart of quantum hardware lies the qubit, the quantum analog of classical bits. Qubits encode information in quantum states that can represent both 0 and 1 simultaneously, a phenomenon known as superposition. Qubits are the fundamental building blocks of quantum processors and serve as the carriers of quantum information. Various physical systems, such as superconducting circuits, trapped ions, and photons, can be used to realize qubits.

Superconducting qubits are among the most widely employed qubit technologies. These qubits are typically fabricated from superconducting materials and require ultra-cold temperatures to operate. Superconducting qubits exhibit long coherence times and can be precisely controlled using microwave pulses. They serve as the basis for many cloud-based quantum computing platforms, including those provided by IBM and Google.

Trapped ions offer another robust qubit technology. In this approach, qubits are encoded in the internal energy levels of individual ions, typically ytterbium or calcium ions. Laser beams are used to manipulate the ions' quantum states, allowing for the execution of quantum gates. Trapped-ion systems are known for their long coherence times and high gate fidelities, making them promising candidates for quantum computing applications.

Photonic qubits, encoded in the states of photons, are central to photonic quantum processors. Photons are inherently immune to decoherence from their environment and can be easily manipulated using optical components such as beam splitters and phase shifters. Photonic qubits are well-suited for tasks like quantum communication and quantum cryptography, where the transmission of quantum information over long distances is crucial.

Topological qubits represent a cutting-edge qubit technology with the potential for robust fault-tolerant quantum computation. Topological qubits are anyons that exhibit non-Abelian statistics, making them highly resistant to errors. However, realizing topological qubits and creating the necessary hardware is an ongoing challenge, and this technology is still in the research phase.

Beyond qubits, quantum processors require precise control electronics and microwave components to manipulate and read out qubit states. These control systems ensure that quantum gates are applied accurately and that qubits are measured with high fidelity. Cryogenic cooling systems are essential to maintain qubits at ultra-cold temperatures, allowing them to maintain their coherence.

Quantum processors also incorporate microwave resonators and cavities, which are used to couple qubits and enable their interactions. These resonators are engineered to operate at specific frequencies and facilitate the controlled manipulation of qubits through microwave pulses.

To achieve scalability, quantum processors often adopt a modular architecture, where multiple qubits are interconnected in a controlled and coherent manner. Quantum interconnects, such as superconducting wires or optical fibers, facilitate qubit-qubit interactions and enable the implementation of multi-qubit gates.

Control electronics play a vital role in quantum hardware, providing the means to precisely apply microwave pulses and control the quantum operations performed on qubits. These electronics require advanced signal generation and measurement capabilities to ensure the accuracy and reliability of quantum gates.

Quantum processors also rely on specialized error correction codes to mitigate the effects of quantum errors. Quantum error correction is an essential component of quantum hardware, as qubits are susceptible to decoherence and gate imperfections. Error correction codes, such as the surface code and the Steane code, introduce redundancy into quantum information, allowing for the detection and correction of errors.

To enable error correction, quantum processors are equipped with error syndrome measurement systems. These systems continuously monitor the state of qubits during quantum computations and detect errors when they occur. The syndromes measured by these systems are crucial for identifying and rectifying errors through error correction codes.

Quantum processors often feature cryogenic systems that maintain qubits at ultra-cold temperatures near absolute zero. Cryogenic cooling is essential for preserving qubit coherence and ensuring stable quantum operations. Sophisticated cryogenic systems are employed to reach and maintain these extreme temperatures.

Quantum processors require precision and scalability in their control and measurement systems. To achieve this, they often integrate custom-designed control electronics and signal processing hardware. These components enable the execution of quantum algorithms with high accuracy and reliability.

Quantum hardware development is a dynamic field, with ongoing efforts to improve qubit quality, gate fidelities, and error correction capabilities. Researchers and engineers are continually pushing the boundaries of quantum hardware to create more powerful and reliable quantum processors.

The pursuit of quantum advantage, where quantum computers outperform classical computers in specific tasks, drives the development of quantum hardware. Quantum hardware components are instrumental in achieving quantum advantage, enabling the execution of quantum algorithms with exponential speedup over classical counterparts.

Quantum hardware is made more accessible through cloud-based quantum computing platforms, allowing researchers and developers worldwide to access and experiment with quantum processors. These platforms offer a range of quantum hardware configurations, from small-scale processors suitable for educational purposes to state-of-the-art quantum computers designed for cutting-edge research and applications.

In summary, quantum hardware components form the bedrock of quantum processors and quantum computers, enabling the manipulation and measurement of qubits to perform quantum computations. Qubits, control electronics, cryogenic systems, and error correction codes are all integral elements of quantum hardware. As quantum technology continues to advance, the development of increasingly powerful and reliable quantum hardware is essential for realizing the full potential of quantum computing in various domains, from cryptography and materials science to optimization and artificial intelligence.

Chapter 9: Quantum Simulations and Cryptography

Embarking on a journey into the realm of quantum computing, one encounters the fascinating world of quantum simulations, where the power of quantum mechanics is harnessed to unravel the mysteries of complex systems. Quantum simulations represent a groundbreaking approach to understanding and predicting the behavior of intricate systems that defy classical computational methods. These simulations hold the promise of revolutionizing fields as diverse as materials science, drug discovery, and fundamental physics by providing insights and solutions that were previously unattainable.

At the heart of quantum simulations are quantum computers, which leverage the principles of superposition and entanglement to manipulate and analyze quantum states. Unlike classical computers, which rely on bits as binary units of information, quantum computers employ qubits as the fundamental building blocks. Qubits can exist in multiple states simultaneously, allowing quantum computers to explore a vast solution space in parallel, a capability that is crucial for simulating complex systems.

Quantum simulations offer a powerful tool for studying quantum systems themselves, where the behavior of particles at the quantum level defies classical intuition. Quantum chemistry is one of the prime examples of a field that benefits from quantum simulations. Understanding the electronic structure and dynamics of molecules is essential for drug discovery, materials design, and chemical reactions. Quantum computers can provide highly accurate simulations of molecular systems, enabling the discovery of new drugs and materials with unprecedented efficiency.

Beyond quantum chemistry, condensed matter physics is another area where quantum simulations excel. The behavior of electrons in solids, such as superconductors or exotic materials, is inherently quantum mechanical. Simulating these systems on classical computers becomes intractable as the number of particles increases. Quantum simulations, however, can model the quantum interactions among particles accurately, paving the way for the discovery of novel materials with exceptional properties.

The simulation of quantum systems extends to the realm of fundamental physics, where researchers seek to explore the mysteries of the universe at the smallest scales. Quantum field theories and lattice gauge theories, which describe the behavior of subatomic particles and the strong force, are notoriously challenging to compute using classical methods. Quantum simulations offer a path to gaining deeper insights into the behavior of particles and forces at the quantum level, potentially unlocking new physics beyond the Standard Model.

Quantum simulations are not limited to the quantum realm; they also find applications in simulating classical systems with complex interactions. For example, simulating the behavior of molecules and materials under extreme conditions, such as high pressure or temperature, is critical for understanding phenomena like superconductivity and phase transitions. Quantum computers can provide accurate simulations of these systems, enabling researchers to explore their properties and potential applications.

Monte Carlo simulations, a powerful technique for modeling complex systems in various fields, benefit significantly from quantum computing. Quantum Monte Carlo methods, which utilize quantum algorithms to simulate the behavior of particles and interactions, can provide more efficient and accurate simulations than classical counterparts. This has implications for applications ranging from financial modeling to simulating biological systems.

Quantum simulations also hold promise for solving optimization problems, where the goal is to find the best solution among a *vast number of possibilities. Many real-world optimization problems, such as route planning and resource allocation, are computationally challenging. Quantum algorithms, such as the Quantum Approximate Optimization Algorithm (QAOA), can be employed to find near-optimal solutions efficiently. This has implications for industries like logistics, finance, and supply chain management.*

Quantum simulations are not without challenges. Building and maintaining quantum computers with a sufficient number of qubits and low error rates is a significant technical hurdle. Error mitigation techniques and quantum error correction codes are essential to ensure the accuracy of quantum simulations. Additionally, mapping the behavior of complex systems onto a quantum computer's architecture requires careful design and optimization.

Hybrid quantum-classical algorithms represent another avenue for quantum simulations. In these approaches, classical computers collaborate with quantum processors to perform simulations. Quantum computers handle the most challenging quantum aspects of the simulation, while classical computers manage the rest. Hybrid algorithms enable researchers to harness the power of quantum computing while mitigating the limitations of current quantum hardware.

Quantum simulators, dedicated quantum devices designed for specific simulations, are also emerging as a practical approach. These devices are tailored to simulate particular quantum systems, allowing for more efficient and accurate simulations. For example, quantum annealers, such as those developed by D-Wave Systems, are specialized quantum simulators optimized for solving optimization problems.

The field of quantum simulations is multidisciplinary, involving expertise in quantum physics, quantum algorithms, and domain-specific knowledge. Quantum software development plays a crucial role in designing and implementing quantum algorithms for simulations. Quantum programming languages, such as Qiskit and Cirq, provide tools and libraries for quantum simulations.

Quantum simulations are poised to disrupt various industries and fields. In drug discovery, for instance, quantum simulations can accelerate the search for new pharmaceutical compounds and predict their behavior with high precision. This has the potential to revolutionize drug development and reduce the time and cost associated with bringing new drugs to market.

Materials science stands to benefit from quantum simulations by enabling the discovery of materials with tailored properties for specific applications, such as energy storage, electronics, and quantum computing itself. Quantum simulations can expedite the development of materials with exceptional properties, such as high-temperature superconductors or advanced catalysts.

Climate modeling is another area where quantum simulations can make a significant impact. Simulating the behavior of molecules and particles in the Earth's atmosphere and oceans with quantum accuracy can lead to more accurate climate models and predictions. This, in turn, can inform policies and strategies for mitigating climate change.

In finance and optimization, quantum simulations have the potential to revolutionize portfolio optimization, risk assessment, and complex financial modeling. Quantum algorithms can solve large-scale optimization problems with unprecedented efficiency, offering a competitive advantage in the financial industry.

As quantum technology continues to advance, quantum simulations will play an increasingly prominent role in scientific research, engineering, and industry. The pursuit of quantum advantage, where quantum computers outperform classical computers in specific tasks, is a driving force behind quantum simulation research and development.

In summary, quantum simulations represent a transformative approach to understanding and solving complex systems that elude classical computational methods. Quantum computers, with their ability to harness the power of quantum mechanics, open new frontiers in simulating quantum and classical systems accurately and efficiently. From quantum chemistry and condensed matter physics to optimization and climate modeling, quantum simulations have the potential to revolutionize numerous fields, offering insights and solutions that were once beyond reach. The ongoing development of quantum hardware and algorithms promises to unlock the full potential of quantum simulations, paving the way for groundbreaking discoveries and applications.

Embarking on a journey into the realm of quantum cryptography, one delves into a world of secure communication protocols that harness the unique properties of quantum mechanics to protect information from eavesdropping and ensure the utmost confidentiality. Quantum cryptography protocols represent a revolutionary approach to securing communication channels, offering unprecedented levels of security that rely on the fundamental principles of quantum physics.

In the realm of classical cryptography, security often hinges on the computational difficulty of solving mathematical problems, such as factorization or discrete logarithms. However, the advent of quantum computers threatens the foundations of classical cryptography by potentially rendering these problems solvable in polynomial time.

Quantum cryptography, on the other hand, leverages the principles of quantum mechanics to provide a level of security that is theoretically unbreakable, even by quantum computers. The cornerstone of quantum cryptography lies in the use of quantum key distribution (QKD) protocols, which enable two parties to establish a shared secret key for secure communication.

One of the earliest and most renowned QKD protocols is the BB84 protocol, developed by Charles Bennett and Gilles Brassard in 1984. The BB84 protocol relies on the properties of quantum bits or qubits to ensure the security of key distribution. In BB84, the sender, often referred to as Alice, encodes a secret key as a sequence of qubits and transmits them to the receiver, often referred to as Bob, through a quantum channel.

The magic of quantum cryptography protocols like BB84 lies in the properties of qubits. In the quantum world, observing a qubit inherently disturbs its state, a phenomenon known as the observer effect. This means that any eavesdropper, often referred to as Eve, attempting to intercept the qubits and measure them to gain knowledge about the key would inevitably introduce errors into the transmission, revealing her presence.

The BB84 protocol incorporates quantum properties such as superposition and entanglement. Alice randomly prepares qubits in one of four states, which correspond to two orthogonal bases. She transmits these qubits to Bob, who randomly chooses one of two measurement bases for each received qubit. Bob's measurement results are then used to establish the shared secret key, while errors caused by Eve's potential interference can be detected through statistical analysis.

While BB84 and other QKD protocols provide a powerful foundation for secure key distribution, they are not without practical challenges. Implementing QKD in real-world scenarios involves overcoming issues related to channel losses, noise, and the need for specialized hardware. Quantum key distribution is typically deployed in point-to-point scenarios over relatively short distances, making it well-suited for securing critical communications, such as those in government and finance.

Beyond BB84, other QKD protocols have been developed to address specific challenges and requirements. The E91 protocol, proposed by Artur Ekert in 1991, utilizes entangled qubits to enhance the security of quantum key distribution. Entangled qubits exhibit correlations that are stronger than those achievable with classical systems, offering a higher level of security. The measurement-device-independent QKD (MDI-QKD) protocol takes security to another level by eliminating potential vulnerabilities associated with the measurement devices used by Bob. In MDI-QKD, the security of the key distribution is guaranteed even if Eve has access to Bob's measurement apparatus. Quantum cryptography protocols extend beyond QKD to encompass other aspects of secure communication. Quantum secure direct communication (QSDC) protocols enable two parties to communicate directly without first establishing a shared key. In QSDC, the sender encodes the message using quantum states and transmits it to the receiver, who can decode the message using quantum measurements.

Quantum key distribution can also be combined with classical encryption techniques to enhance security. Hybrid encryption schemes, which integrate quantum key distribution with classical encryption algorithms, provide an additional layer of protection for data transmission. Quantum keys are used to securely exchange classical encryption keys, ensuring confidentiality and integrity.

The development and deployment of quantum cryptography protocols are influenced by advancements in quantum technology. Quantum key distribution systems have evolved to incorporate quantum repeaters, which extend the range of secure communication over long distances. Quantum repeaters enable the distribution of quantum keys across extensive networks, making quantum cryptography suitable for applications like secure telecommunications.

The field of post-quantum cryptography, which aims to develop cryptographic algorithms that remain secure against quantum attacks, is also closely related to quantum cryptography. Post-quantum cryptographic algorithms are designed to withstand *attacks by quantum computers, ensuring the long-term security of encrypted data.*

Quantum cryptography protocols are not limited to terrestrial applications. Quantum communication technologies have been proposed for secure communication in space, offering a means of safeguarding data transmissions between satellites and ground stations. Quantum key distribution can play a pivotal role in ensuring the security of space-based communications, which are susceptible to interception and interference.

As quantum technology continues to advance, quantum cryptography protocols are poised to play an increasingly significant role in safeguarding sensitive information and enabling secure communication in an era of quantum computing. The unbreakable security provided by these protocols makes them a critical component of the cybersecurity landscape, offering protection against the potential threat posed by quantum computers to classical cryptographic methods.

In summary, quantum cryptography protocols represent a groundbreaking approach to secure communication that harnesses the principles of quantum mechanics to guarantee the confidentiality and integrity of information. Quantum key distribution protocols, such as BB84 and E91, exploit the properties of qubits to thwart eavesdropping attempts. Quantum secure direct communication and hybrid encryption schemes expand the scope of quantum cryptography to address various communication requirements. Advances in quantum technology, such as quantum repeaters and post-quantum cryptography, further enhance the capabilities and resilience of quantum cryptography protocols. In an increasingly interconnected and data-driven world, the deployment of quantum cryptography is poised to play a pivotal role in ensuring the security of sensitive information and communication channels.

Chapter 10: The Future of Quantum Computing

Journeying into the realm of quantum computing, one encounters the concept of quantum supremacy, a milestone that marks the point at which quantum computers surpass classical computers in solving specific tasks. Quantum supremacy represents a pivotal moment in the development of quantum technology, demonstrating the immense computational power that quantum computers can unleash.

The concept of quantum supremacy was introduced by John Preskill in 2012 as a way to describe the potential advantage of quantum computers over classical computers in certain domains. Quantum computers leverage the principles of quantum mechanics, such as superposition and entanglement, to perform calculations at a scale and speed that classical computers cannot match.

To achieve quantum supremacy, a quantum computer must perform a task that is practically impossible for even the most advanced classical supercomputers to complete within a reasonable timeframe. Google's quantum computer, Sycamore, achieved this milestone in 2019 when it successfully completed a computational task in just 200 seconds that would have taken the most powerful classical supercomputer over 10,000 years to solve.

The specific task that Sycamore accomplished was a random quantum circuit sampling problem, designed to demonstrate quantum superiority. This achievement ignited a flurry of excitement and debate in the quantum computing community, as it showcased the potential of quantum computers to outperform classical counterparts on specific problems.

Quantum supremacy, however, does not imply that quantum computers are superior in all computational tasks. Classical computers remain highly efficient for many practical purposes, and quantum computers are still in the early stages of development. Quantum supremacy is more about demonstrating the potential of quantum technology rather than rendering classical computers obsolete.

Beyond quantum supremacy lies the pursuit of practical quantum advantage, where quantum computers can outperform classical computers in tasks that have real-world applications. Practical quantum advantage has the potential to revolutionize industries such as cryptography, materials science, optimization, and artificial intelligence.

Quantum cryptography, for example, can benefit from quantum computers by leveraging their computational power to break classical encryption methods. This poses a significant challenge to the cybersecurity landscape, but it also motivates the development of post-quantum cryptographic algorithms that are resistant to quantum attacks.

Materials science stands to gain from quantum computing through the simulation of complex quantum systems, leading to the discovery of new materials with unique properties. Quantum computers can expedite the development of materials for applications like energy storage, electronics, and quantum information processing.

Optimization problems, which are prevalent in logistics, finance, and supply chain management, can be solved more efficiently using quantum algorithms. Quantum computers can provide near-optimal solutions to complex optimization problems, offering a competitive advantage in these industries.

Machine learning and artificial intelligence can also benefit from quantum computing. Quantum machine learning *algorithms have the potential to accelerate tasks like data analysis and pattern recognition. Quantum neural networks and quantum-enhanced algorithms may unlock new possibilities in the field of artificial intelligence.*

Quantum chemistry and drug discovery are domains where quantum computers can make a significant impact. Simulating the behavior of molecules and predicting chemical reactions with high accuracy can lead to the discovery of new drugs and materials, ultimately improving healthcare and pharmaceutical development.

Climate modeling and environmental research can benefit from quantum simulations, enabling more accurate predictions of climate patterns and the effects of climate change. Quantum simulations can inform policies and strategies for mitigating the impacts of global warming.

The pursuit of practical quantum advantage is not without challenges. Quantum hardware development is a complex and ongoing endeavor. Quantum processors must be engineered with a sufficient number of qubits and low error rates to tackle real-world problems effectively.

Quantum error correction is crucial for building reliable and fault-tolerant quantum computers. Error correction codes and techniques are essential to mitigate the effects of noise and errors in quantum computations.

Hybrid quantum-classical algorithms are emerging as a practical approach to harnessing the power of quantum computers while mitigating their limitations. In hybrid algorithms, quantum computers collaborate with classical computers to solve complex problems efficiently.

Quantum software development is a critical component of achieving practical quantum advantage. Quantum programming languages and libraries, such as Qiskit, Cirq, and Quipper, provide tools and frameworks for designing and implementing quantum algorithms.

Quantum algorithms themselves are a subject of ongoing research and optimization. Researchers are continuously exploring new quantum algorithms and improving existing ones to maximize the advantages of quantum computing.

Quantum networking and quantum communication are integral to realizing the full potential of quantum technology. Quantum entanglement and quantum teleportation enable secure and instantaneous transmission of quantum information, paving the way for quantum internet and quantum-enhanced communication protocols.

In the pursuit of practical quantum advantage, interdisciplinary collaboration is essential. Quantum technology spans multiple domains, including physics, computer science, materials science, and engineering. Researchers and experts from various fields must work together to push the boundaries of quantum computing and its applications.

Quantum education and workforce development play a vital role in advancing quantum technology. Training the next generation of quantum scientists, engineers, and researchers is essential for driving innovation and progress in the field.

Quantum ethics and policy considerations are also important as quantum technology evolves. The ethical implications of quantum computing, such as the potential impact on privacy and security, require careful consideration and regulatory frameworks.

Quantum technology is poised to transform industries and solve problems that are currently beyond the reach of classical computers. The journey from quantum supremacy to practical quantum advantage is an exciting and dynamic one, filled with challenges and opportunities.

In summary, quantum supremacy represents a significant milestone in the development of quantum technology, showcasing the computational power of quantum computers. Beyond quantum supremacy lies the pursuit of practical quantum advantage, where quantum computers can revolutionize industries and solve real-world problems. Quantum cryptography, materials science, optimization, machine learning, and climate modeling are just a few areas poised to benefit from quantum computing. Overcoming technical challenges, advancing quantum algorithms, and fostering interdisciplinary collaboration are essential for realizing the full potential of quantum technology. The journey into the quantum era promises to usher in a new era of innovation and discovery, with quantum computing at its forefront.

As we delve into the world of emerging quantum technologies, we find ourselves at the forefront of a scientific and technological revolution that promises to reshape our understanding of the universe and revolutionize various industries. Quantum technologies harness the profound and often counterintuitive principles of quantum mechanics to unlock new capabilities and applications that were once deemed impossible.

Quantum computing stands as one of the most prominent and transformative emerging quantum technologies. Quantum computers leverage the properties of quantum bits or qubits, which can exist in multiple states simultaneously due to superposition. This enables quantum computers to perform calculations at an unprecedented speed and scale, potentially revolutionizing fields such as cryptography, optimization, materials science, and drug discovery.

Quantum cryptography, another pivotal emerging quantum technology, promises secure communication in an era when classical encryption methods may become vulnerable to quantum attacks. Quantum key distribution (QKD) protocols, which rely on the principles of quantum mechanics to *establish secure keys, offer an unbreakable level of security that can protect sensitive data from eavesdropping and interception.*

Quantum sensing and metrology represent another frontier in emerging quantum technologies. Quantum sensors, such as atomic clocks and magnetometers, exploit quantum properties to achieve unprecedented precision in measuring time, magnetic fields, and gravitational forces. These technologies have far-reaching applications in fields like navigation, geophysics, and fundamental physics experiments.

Quantum communication networks are poised to revolutionize the way we transmit and process information. Quantum entanglement and quantum teleportation enable secure and instantaneous communication over long distances, paving the way for quantum-enhanced internet and communication protocols that protect data from interception.

Quantum imaging and microscopy, utilizing entangled photon pairs and quantum states of light, offer the potential for ultra-sensitive and high-resolution imaging techniques. These technologies have applications in medical imaging, biological research, and materials characterization.

Quantum simulators, dedicated quantum devices designed to mimic and understand complex quantum systems, are emerging as a powerful tool for studying quantum behavior and solving problems in quantum chemistry, condensed matter physics, and fundamental research.

Quantum machine learning, a fusion of quantum computing and artificial intelligence, holds the promise of accelerating tasks like data analysis, pattern recognition, and optimization. Quantum neural networks and quantum-enhanced algorithms may lead to breakthroughs in machine learning applications.

Quantum hardware components, such as quantum processors and quantum memory, continue to advance, paving the way for the practical realization of quantum technologies. Quantum error correction codes and quantum hardware innovations are essential for building reliable and fault-tolerant quantum computers.

Quantum software development is a critical aspect of emerging quantum technologies. Quantum programming languages, libraries, and algorithms provide the tools and frameworks needed to design and implement quantum solutions in various domains.

Quantum education and workforce development play a pivotal role in the adoption and advancement of quantum technologies. Training the next generation of quantum scientists, engineers, and researchers is essential for driving innovation and progress in the field.

Quantum ethics and policy considerations are integral as quantum technologies evolve. Addressing ethical concerns, privacy issues, and regulatory frameworks for quantum technologies is essential for responsible development and deployment.

Quantum computing, in particular, is poised to disrupt industries and solve complex problems that have remained unsolvable using classical computers. Quantum algorithms, such as Shor's algorithm and Grover's algorithm, have the potential to break classical encryption methods and revolutionize cryptography.

Quantum optimization algorithms can find near-optimal solutions to complex problems in logistics, finance, and supply chain management. Quantum annealers, like those developed by D-Wave Systems, are specialized quantum devices designed for solving optimization problems.

Quantum chemistry and materials science benefit from quantum simulations, enabling researchers to understand

and design molecules and materials with tailored properties for applications in drug discovery, materials design, and energy storage.

Climate modeling and environmental research can leverage quantum simulations to enhance our understanding of climate patterns and the effects of climate change. Quantum simulations offer the potential to inform policies and strategies for mitigating the impacts of global warming.

In quantum cryptography, quantum key distribution (QKD) protocols like BB84 and E91 provide secure communication channels that are theoretically invulnerable to quantum attacks. These protocols are critical for safeguarding sensitive data in a quantum-powered world.

Quantum sensors, such as atomic clocks, are at the heart of emerging technologies like global positioning systems (GPS). The precision and accuracy of quantum sensors enable more reliable navigation, timing, and geolocation services.

Quantum communication networks are a vital component of quantum technologies. Quantum entanglement enables secure communication channels that can protect data from interception, making them crucial for industries like finance, healthcare, and defense.

Quantum imaging and microscopy techniques offer new insights into the microscopic world. Quantum-enhanced imaging can revolutionize medical diagnostics, materials science, and biological research.

Quantum machine learning algorithms have the potential to transform industries by accelerating tasks like data analysis and optimization. Quantum neural networks and quantum-enhanced algorithms may lead to innovative applications in finance, healthcare, and autonomous systems.

Quantum hardware development is essential for realizing the full potential of quantum technologies. Advancements in *qubit quality, gate fidelities, and error correction capabilities are driving progress in quantum computing.*

Quantum software development is a dynamic field, with quantum programming languages like Qiskit and Cirq providing the tools needed to design and implement quantum algorithms. Quantum libraries and frameworks facilitate quantum software development in various domains.

Quantum education and workforce development programs are essential for cultivating the talent and expertise needed to advance quantum technologies. Training the next generation of quantum scientists and engineers is critical for driving innovation and progress in the field.

Quantum ethics and policy considerations are crucial as quantum technologies continue to evolve. Addressing ethical concerns, privacy issues, and regulatory frameworks for quantum technologies is essential for responsible development and deployment.

In summary, emerging quantum technologies hold the promise of revolutionizing industries, solving complex problems, and unlocking new capabilities that were once the stuff of science fiction. Quantum computing, quantum cryptography, quantum sensing, and quantum communication networks are just a few examples of the transformative potential of quantum technologies. As research and development in the field continue to advance, we stand at the threshold of a quantum-powered future that will reshape the way we live, work, and interact with the world around us.

Top of Form

BOOK 2
MASTERING QUANTUM COMPUTING
A COMPREHENSIVE GUIDE FOR INTERMEDIATE LEARNERS
ROB BOTWRIGHT

Chapter 1: Reviewing Quantum Basics for Intermediate Learners

In the realm of quantum mechanics, the fundamental principles governing the behavior of particles on the smallest scales of the universe, we encounter a fascinating and perplexing world that challenges our classical intuitions. This branch of physics, born in the early 20th century, revolutionized our understanding of the physical universe by introducing concepts that seem counterintuitive yet underpin the very fabric of reality.

One of the central tenets of quantum mechanics is the wave-particle duality, a concept that defies classical notions of matter and energy. According to this principle, particles, such as electrons and photons, can exhibit both wave-like and particle-like properties depending on how they are observed or measured. This dual nature gives rise to phenomena like interference, where waves can cancel each other out or reinforce one another, creating patterns of light and darkness that seem to defy classical logic.

Quantum mechanics introduces another intriguing concept: superposition. It suggests that particles can exist in multiple states simultaneously, blending different probabilities of being in various states until observed or measured, at which point they "collapse" into a single definite state. This inherent uncertainty challenges our classical determinism and implies that the universe operates with inherent randomness at its core.

The famous Schrödinger's cat thought experiment illustrates the concept of superposition. Imagine a cat enclosed in a sealed box with a radioactive atom and a vial of poison. In this quantum scenario, the cat is simultaneously alive and dead until someone opens the box to observe its fate, *collapsing the superposition into one outcome or the other. This paradoxical idea forces us to reconsider our understanding of reality.*

Quantum mechanics also introduces the concept of entanglement, where particles become correlated in such a way that the state of one particle instantly affects the state of another, regardless of the distance separating them. Einstein famously referred to this phenomenon as "spooky action at a distance." Entanglement challenges our classical notions of locality and suggests that information can be transmitted instantaneously between entangled particles, raising questions about the nature of causality itself.

To describe the quantum world, physicists developed a mathematical framework called the wavefunction, often denoted by the Greek letter Ψ (psi). The wavefunction encapsulates all the information about a quantum system, representing the probabilities of finding particles in different states. When squared, the absolute value of the wavefunction gives the probability density, offering insight into the likelihood of finding a particle in a particular position or state.

The Schrödinger equation, a fundamental equation in quantum mechanics, describes how the wavefunction evolves over time. This equation is analogous to Newton's second law in classical mechanics but accounts for the probabilistic nature of quantum systems. Solving the Schrödinger equation allows physicists to predict the behavior of particles on the quantum level, providing a powerful tool for understanding the microscopic world.

In addition to the wavefunction, another critical concept in quantum mechanics is the Heisenberg Uncertainty Principle. Proposed by Werner Heisenberg in 1927, this principle asserts that there is a fundamental limit to how precisely we can simultaneously know the position and momentum of a particle. The more accurately we determine one of these properties, the less accurately we can know the other. This inherent uncertainty is a fundamental aspect of the quantum world, challenging our classical notions of determinism.

Quantum mechanics also introduces the concept of quantization, where physical properties, such as energy levels in atoms, are quantized into discrete values. These quantized levels give rise to phenomena like atomic spectra, where atoms emit or absorb light at specific wavelengths corresponding to transitions between energy levels. This quantization of energy levels is a hallmark of the quantum world and has profound implications for understanding the behavior of matter and light.

Another crucial concept in quantum mechanics is tunneling, a phenomenon where particles can penetrate energy barriers that classical physics would consider impenetrable. Tunneling plays a vital role in various physical processes, from the operation of transistors in electronic devices to the fusion of hydrogen nuclei in stars. This phenomenon underscores the importance of quantum effects in the macroscopic world.

Quantum mechanics also has profound implications for the field of cryptography, as it offers the possibility of secure communication through quantum key distribution. Quantum encryption relies on the principles of superposition and entanglement to create a secure channel for transmitting cryptographic keys. The inherent randomness and indeterminacy of quantum systems make it theoretically impossible for eavesdroppers to intercept and decode the keys without detection, providing an unprecedented level of security for information transmission.

Moreover, the development of quantum computers has the potential to revolutionize computing by exploiting the principles of superposition and entanglement to perform complex calculations at speeds unattainable by classical computers. Quantum algorithms, such as Shor's algorithm for factoring large numbers, threaten classical encryption methods and raise questions about the future of cybersecurity.

Quantum mechanics also has practical applications in the development of quantum sensors and metrology tools. Quantum sensors, such as atomic clocks and magnetometers, utilize the precise measurement capabilities of quantum systems to achieve unprecedented levels of accuracy in various fields, from navigation to fundamental physics experiments.

Exploring the world of quantum mechanics challenges our classical intuitions and forces us to confront the inherent probabilistic and indeterministic nature of the universe on the smallest scales. While these concepts may seem counterintuitive, they underpin our understanding of the quantum world and have led to groundbreaking discoveries and technological advancements. As we continue to unravel the mysteries of quantum mechanics, we open doors to new possibilities and reshape our perception of reality itself.

In the world of quantum mechanics, understanding quantum states and operators is fundamental. Quantum states are the core building blocks of this fascinating and counterintuitive branch of physics. These states describe the properties of quantum systems, providing a complete and unique description of a particle's behavior.

A quantum state, often represented as a ket vector ($|\psi\rangle$), encapsulates all the information about a quantum system, including its position, momentum, spin, and any other measurable quantities. This vector exists in a complex vector space, where its components are complex numbers, and its inner product is used to calculate probabilities and outcomes.

Operators are mathematical entities that act on quantum states, transforming them into new states or extracting information about the system. These operators correspond to physical observables like position, momentum, and angular momentum. In quantum mechanics, operators are represented by Hermitian matrices or linear operators that possess unique properties.

One of the most fundamental operators in quantum mechanics is the Hamiltonian operator (H). The Hamiltonian operator represents the total energy of a quantum system and is essential for determining how the system evolves over time. It plays a central role in the time-dependent Schrödinger equation, which describes how quantum states change as a function of time.

The Schrödinger equation, a cornerstone of quantum mechanics, is given by the equation:

$$H|\psi\rangle = i\hbar\, \partial|\psi\rangle / \partial t.$$

Here, H is the Hamiltonian operator, $|\psi\rangle$ is the quantum state, $\hbar$ (h-bar) is the reduced Planck constant, and $\partial/\partial t$ represents the partial derivative with respect to time. This equation describes the time evolution of a quantum state, allowing us to predict how the state changes over time in response to the Hamiltonian operator.

Quantum states can be either stationary states or superpositions of multiple states. Stationary states are eigenstates of the Hamiltonian operator, meaning that when the Hamiltonian operates on these states, they remain unchanged except for a phase factor. These stationary states are associated with definite energy levels and represent stable states in a quantum system.

Superposition is another key concept in quantum mechanics. It allows quantum states to exist as linear combinations of multiple basis states, each with its own probability amplitude. Superposition enables particles to be in multiple states simultaneously, giving rise to phenomena like interference and entanglement.

The probability density of finding a particle in a particular state is given by the squared magnitude of the probability amplitude associated with that state. Measurement outcomes in quantum mechanics are inherently probabilistic, and the Born rule provides a mathematical framework for calculating these probabilities.

Operators in quantum mechanics correspond to physical observables, and their properties play a crucial role in quantum theory. Hermitian operators are self-adjoint, meaning that they are equal to their own adjoints. The adjoint of an operator is obtained by taking its complex conjugate transpose. Hermitian operators have real eigenvalues, which correspond to the possible measurement outcomes of the observable they represent.

The position operator ($\hat{x}$) is a fundamental operator in quantum mechanics that corresponds to the position of a particle along a particular axis. When the position operator acts on a quantum state, it extracts the position information of the particle. The momentum operator ($\hat{p}$), on the other hand, corresponds to the momentum of a particle and is related to the position operator through the Heisenberg Uncertainty Principle.

The Heisenberg Uncertainty Principle is a fundamental concept in quantum mechanics that states that the more precisely we know a particle's position, the less precisely we can know its momentum, and vice versa. This inherent uncertainty is a fundamental limitation in quantum mechanics and challenges our classical notions of determinism.

Spin is another important property of quantum particles. It is associated with intrinsic angular momentum and can take on discrete values, such as up or down. Spin operators, represented as $\hat{S}$, act on quantum states to extract information about a particle's spin, and they obey unique commutation relations.

In addition to position, momentum, and spin, quantum mechanics encompasses various other observables and operators, including angular momentum, energy, and more. Each observable corresponds to a unique operator, and the eigenstates of these operators provide a complete description of the quantum system.

Quantum states can evolve over time, and the time evolution is governed by the Schrödinger equation. This equation describes how the quantum state changes as a function of time and allows us to predict the future behavior of a quantum system based on its initial state and the Hamiltonian operator.

Quantum operators can also evolve over time through a process known as unitary transformation. Unitary operators are those that preserve the inner product of quantum states and are essential for understanding how observables change as a quantum system evolves.

In summary, quantum states and operators are at the heart of quantum mechanics, providing a comprehensive framework for understanding and describing the behavior of quantum systems. Quantum states encapsulate all the information about a system, while operators correspond to physical observables and play a central role in predicting measurement outcomes and understanding the evolution of quantum states over time. These concepts, while often challenging and counterintuitive, are essential for unraveling the mysteries of the quantum world.

Chapter 2: Quantum Gates and Quantum Circuits Revisited

Delving deeper into the fascinating realm of quantum mechanics, we come across the intricate and powerful domain of advanced quantum gate operations. These operations are the fundamental building blocks of quantum computing, enabling us to manipulate and harness the full computational potential of quantum systems.

Quantum gates are analogous to classical logic gates, but they operate on quantum bits or qubits, which can exist in a superposition of states, unlike classical bits that are either 0 or 1. Advanced quantum gate operations allow us to perform complex quantum computations, opening doors to solving problems that were previously beyond the reach of classical computers.

One of the foundational quantum gates is the Pauli-X gate, often referred to as the quantum NOT gate. It flips the state of a qubit, transforming $|0\rangle$ into $|1\rangle$ and vice versa. The Pauli-X gate is a crucial component for creating entangled states and quantum circuits that perform various quantum algorithms.

Another essential quantum gate is the Pauli-Y gate, which introduces a complex phase to the qubit states while flipping them. This gate is pivotal in creating quantum circuits for error correction and fault-tolerant quantum computing.

The Pauli-Z gate, the third member of the Pauli gate family, introduces a phase shift based on the state of the qubit. It doesn't change the basis states $|0\rangle$ and $|1\rangle$ but adds a phase of -1 to $|1\rangle$, making it a vital component for quantum error correction and stabilizer codes.

In addition to the Pauli gates, we have the Hadamard gate, a fundamental gate in quantum computing. The Hadamard gate creates superpositions and transforms $|0\rangle$ into the state $|+\rangle$ (an equal superposition of $|0\rangle$ and $|1\rangle$) and $|1\rangle$ into the state $|-\rangle$ (an equal superposition with a relative phase of π). This gate plays a pivotal role in many quantum algorithms, including the famous quantum algorithm for Grover's search problem.

The Controlled-NOT (CNOT) gate is another fundamental quantum gate. It operates on two qubits and flips the target qubit's state only when the control qubit is in the $|1\rangle$ state. The CNOT gate is essential for creating entanglement and building quantum circuits that perform quantum error correction.

Quantum gates can be combined to form quantum circuits, allowing us to perform a wide range of quantum operations. These circuits are the quantum analogs of classical digital logic circuits, and they enable us to execute quantum algorithms and computations.

Quantum gate operations also include the T gate, which introduces a $\pi/4$ phase shift to the $|1\rangle$ state while leaving $|0\rangle$ unchanged. The T gate is an essential component in many quantum algorithms, especially those involving quantum Fourier transforms.

The Controlled-U gate, where U represents any single-qubit unitary operation, is a versatile gate that allows us to perform conditional operations based on the state of a control qubit. It forms the basis for many quantum algorithms and quantum error correction codes.

The Swap gate is a two-qubit gate that swaps the states of two qubits. It is useful in quantum circuits for reordering qubits and implementing quantum teleportation.

Quantum gates also include the Toffoli gate, a three-qubit gate that performs a controlled-controlled-not operation. It flips the target qubit's state only when both control qubits are in the $|1\rangle$ state. The Toffoli gate is essential for building reversible classical logic circuits and is a building block for many quantum algorithms.

Furthermore, the Fredkin gate, another three-qubit gate, swaps the second and third qubits' states based on the state of the first qubit. It is also known as the controlled-swap gate and is used in reversible computing and quantum information processing.

Quantum gate operations play a pivotal role in quantum error correction, a critical aspect of quantum computing. Error correction codes, such as the surface code, rely on a combination of gates and measurements to detect and correct errors that can occur during quantum computations.

In quantum computing, gates are applied to qubits to manipulate their states and perform quantum operations. These gates, when combined into quantum circuits, enable us to execute quantum algorithms that take advantage of the inherent parallelism and superposition properties of quantum systems.

The beauty of quantum gates lies in their ability to create complex entangled states and execute operations that are exponentially faster than classical counterparts for specific problems. Quantum gates are the key to harnessing the full potential of quantum computers, which have the potential to revolutionize fields such as cryptography, materials science, optimization, and more.

Advanced quantum gate operations pave the way for quantum supremacy, a point where quantum computers can outperform classical computers on certain tasks. Achieving quantum supremacy requires not only the development of powerful quantum gates but also addressing the formidable challenges of quantum error correction, decoherence, and noise in quantum systems.

In summary, advanced quantum gate operations are at the heart of quantum computing, enabling us to manipulate and *harness the unique properties of quantum systems. These gates, along with the principles of superposition and entanglement, hold the promise of ushering in a new era of computing and solving complex problems that were previously intractable. The exploration and development of quantum gates continue to push the boundaries of what is possible in the realm of quantum mechanics, opening doors to exciting opportunities and transformative technologies.*

Designing complex quantum circuits is a fascinating journey into the world of quantum computation and information processing, where the manipulation of quantum bits, or qubits, leads to the realization of powerful quantum algorithms. Quantum circuits are analogous to classical digital logic circuits but harness the unique properties of quantum mechanics to perform computations in ways that classical computers cannot.

At the core of quantum circuits are quantum gates, the elementary operations that manipulate qubits. These gates include the Pauli-X, Pauli-Y, Pauli-Z, Hadamard, and Controlled-NOT (CNOT) gates, among others, each with specific functions and characteristics that enable quantum computations. The choice and arrangement of these gates determine the behavior of the quantum circuit.

Quantum gates are often represented as matrices or unitary operators, acting on the quantum states of qubits. The mathematics behind these gates is rooted in linear algebra and quantum mechanics, and their proper combination allows for the construction of complex quantum algorithms.

Designing complex quantum circuits requires careful consideration of the problem at hand and the desired outcomes. Quantum algorithms, such as Shor's algorithm for integer factorization or Grover's algorithm for unstructured search, are designed by strategically arranging quantum gates to exploit quantum parallelism and interference effects.

One of the key principles in designing quantum circuits is quantum parallelism, which allows quantum computers to explore multiple possibilities simultaneously. This property is a result of superposition, where qubits can exist in multiple states at once. Quantum gates are used to create and manipulate these superpositions, enabling quantum computers to process vast amounts of information in parallel.

Interference is another essential concept in quantum circuit design. It occurs when the probability amplitudes of different quantum states interfere constructively or destructively, leading to specific outcomes when measurements are made. Quantum algorithms are designed to carefully control interference to amplify the probability of obtaining the correct answer while suppressing incorrect ones.

Quantum circuit design involves breaking down a quantum algorithm into a sequence of quantum gates, each of which performs a specific operation on the qubits. The order and arrangement of these gates are critical to the algorithm's success. Quantum compilers and software tools assist in automating the translation of high-level quantum algorithms into low-level quantum circuits.

One of the fundamental aspects of quantum circuit design is quantum entanglement. Entanglement occurs when qubits become correlated in such a way that the state of one qubit depends on the state of another, regardless of the distance between them. Entanglement plays a crucial role in quantum computing, enabling the creation of powerful quantum circuits that exploit non-classical correlations to solve complex problems efficiently.

Quantum error correction is an essential consideration in the design of complex quantum circuits. Quantum computers are highly susceptible to errors due to factors such as decoherence and noise. Quantum error correction codes and techniques are employed to detect and correct errors, ensuring the reliability of quantum computations.

The concept of gate fidelity is crucial in quantum circuit design. Fidelity measures how closely a quantum gate operation approximates the ideal quantum gate. High gate fidelity is essential for maintaining the accuracy of quantum computations. Improvements in gate fidelity are a subject of ongoing research in the field of quantum computing.

Quantum circuit design is not limited to algorithmic tasks alone. It also plays a significant role in quantum simulations, where quantum circuits are used to model and study the behavior of complex quantum systems, such as molecules or materials. Quantum simulations have the potential to revolutionize fields like chemistry and materials science by providing insights into systems that are too challenging for classical computers to simulate accurately.

Quantum circuit design is an interdisciplinary field that combines physics, computer science, and mathematics. Quantum algorithms and quantum gates are developed with a deep understanding of the underlying quantum principles, and their implementation relies on the advancement of quantum hardware and technologies.

Hardware constraints, such as qubit connectivity and gate error rates, influence the design of quantum circuits. Quantum circuit designers must work within these constraints to achieve the desired computational outcomes. Quantum hardware improvements, such as the development of fault-tolerant quantum computers, are ongoing efforts aimed at expanding the capabilities of quantum circuit design.

Quantum circuit design also considers the practical aspects of quantum computation, including the preparation and measurement of qubits. Quantum states must be carefully initialized, and measurement outcomes must be interpreted correctly to obtain meaningful results. Quantum circuit designers work on optimizing these processes for efficient quantum computation.

Quantum circuit design is not limited to academic research but also extends to industry applications. Companies and organizations are exploring quantum computing for various tasks, including optimization, cryptography, machine learning, and drug discovery. Quantum circuit designers play a crucial role in developing practical quantum applications that leverage the power of quantum computation.

In summary, designing complex quantum circuits is a multidisciplinary endeavor that combines quantum physics, mathematics, and computer science. Quantum circuits are constructed from quantum gates, each with its unique functionality, to perform quantum algorithms and simulations. Quantum circuit design leverages principles like quantum parallelism, interference, and entanglement to solve complex problems efficiently. It also addresses challenges such as quantum error correction, gate fidelity, and hardware constraints. As quantum technology continues to advance, the field of quantum circuit design holds the promise of unlocking new possibilities and transforming industries across the globe.

Chapter 3: Advanced Topics in Quantum Algorithms

Quantum algorithm optimization is a critical aspect of harnessing the full potential of quantum computing for practical applications. Optimizing quantum algorithms involves enhancing their efficiency, reducing resource requirements, and improving their performance compared to classical counterparts. Quantum computers promise exponential speedup for specific problems, but realizing this potential often requires careful algorithmic design and optimization. One of the primary goals of quantum algorithm optimization is to minimize the number of quantum gates required to solve a given problem. Reducing gate count is essential because quantum gates are susceptible to errors and decoherence, which can limit the accuracy and reliability of quantum computations. Quantum algorithms are typically expressed as sequences of quantum gates that manipulate qubits to perform specific tasks. The optimization process involves finding ways to simplify these gate sequences while preserving the algorithm's correctness and accuracy. Quantum algorithms are designed to exploit quantum properties like superposition and entanglement to achieve computational advantages. Optimization techniques aim to maximize the utilization of these quantum features to minimize the resources needed for a computation. Quantum algorithm optimization is a challenging endeavor that requires a deep understanding of both quantum mechanics and computational complexity theory. Researchers and quantum algorithm designers work on developing new quantum algorithms and optimizing existing ones to tackle various problems efficiently. Quantum algorithms fall into different categories, such as quantum simulation, quantum search, quantum cryptography, and quantum machine learning. Each category has its unique optimization challenges and opportunities. Quantum simulation algorithms aim to simulate quantum systems efficiently, providing insights into quantum chemistry, material science, and other fields. Optimizing quantum simulation algorithms involves finding ways to represent quantum systems with fewer qubits and gates while maintaining accuracy. Quantum search algorithms, like Grover's algorithm, are designed to search unstructured databases faster than classical algorithms. Optimizing quantum search algorithms involves minimizing the number of queries to the database and maximizing the probability of finding the correct solution. Quantum cryptography algorithms, such as quantum key distribution protocols, aim to provide secure communication channels based on the principles of quantum mechanics. Optimization in quantum cryptography focuses on improving the efficiency and security of these protocols against various types of attacks. Quantum machine learning algorithms leverage quantum computing to accelerate tasks like data classification, optimization, and pattern recognition. Optimization in quantum machine learning involves enhancing the speed and accuracy of quantum-enhanced machine learning models. Quantum algorithm optimization also considers the hardware constraints of quantum computers, such as qubit connectivity and gate error rates. Optimal gate sequences may vary depending on the specific quantum hardware being used, and optimization must take these factors into account. Quantum error correction is another critical aspect of quantum algorithm optimization. Quantum computers are inherently susceptible to errors, and error-correcting codes and techniques are employed to mitigate the impact of these errors. Optimizing quantum error correction involves finding efficient codes and *error mitigation strategies that minimize resource overhead while maintaining the integrity of quantum computations. In many cases, quantum algorithm optimization requires finding a trade-off between gate count, error correction, and computational accuracy. Quantum algorithm designers must strike a balance to achieve practical quantum algorithms that outperform classical counterparts. Quantum algorithms often involve complex mathematical operations and quantum gates that manipulate qubits according to specific rules. Optimization may involve finding shortcuts or more efficient ways to perform these operations. Quantum algorithm optimization can also benefit from the use of hybrid quantum-classical algorithms. In hybrid algorithms, certain parts of the computation are offloaded to classical computers, allowing quantum computers to focus on tasks where they excel. Hybrid algorithms leverage the strengths of both quantum and classical computing to achieve improved performance. Quantum algorithm optimization is an ongoing research area, with continuous efforts to discover new algorithms and refine existing ones. The development of quantum hardware, such as quantum processors and quantum annealers, also drives the need for algorithmic optimization. Quantum annealers, for example, are specialized quantum devices that excel at solving optimization problems. Optimizing quantum annealing algorithms involves mapping complex problems onto the hardware architecture of these devices effectively. Quantum algorithm optimization extends beyond theory and research; it has practical applications across various domains. Industries and organizations are exploring the potential of quantum computing to solve complex optimization problems in fields like logistics, finance, and drug discovery. For example, quantum algorithms can optimize supply chain routes, portfolio management, and molecular simulations.*

Optimization problems that are difficult or computationally infeasible for classical computers can be addressed with quantum algorithms. Quantum-inspired optimization algorithms are a class of algorithms that take inspiration from quantum principles to solve optimization problems using classical hardware. These algorithms aim to bridge the gap between classical and quantum optimization techniques and offer computational advantages. Quantum-inspired algorithms leverage concepts like quantum annealing and quantum adiabatic evolution to explore solution spaces more efficiently. Quantum-inspired optimization has the advantage of being accessible on classical hardware while still benefiting from quantum-inspired strategies. Quantum computing companies and startups are actively developing quantum software and algorithms to address real-world optimization challenges. They work on optimizing algorithms for their quantum hardware, ensuring efficient resource utilization and achieving competitive advantages. Quantum algorithm optimization is not limited to a single approach or methodology; it encompasses a diverse range of techniques and strategies. Researchers explore quantum variational algorithms, quantum annealing approaches, and quantum adiabatic optimization, among others. Variational quantum algorithms, such as the variational quantum eigensolver (VQE), are designed to find approximate solutions to complex optimization problems. These algorithms involve parameterized quantum circuits that are optimized to minimize an objective function, providing practical solutions to optimization challenges. Quantum annealing is a specialized optimization technique that leverages quantum effects to search for the global minimum of an objective function. Quantum annealers are designed to tackle optimization problems by finding the lowest-energy state of a quantum system that represents the problem space.

Quantum adiabatic optimization is another approach that aims to find the ground state of a quantum system, which corresponds to the optimal solution of an optimization problem. Adiabatic quantum computing involves gradually transforming a quantum system from an initial state to the desired ground state while maintaining quantum coherence. Quantum algorithm optimization is a dynamic field with the potential to revolutionize industries and solve complex problems. As quantum hardware continues to advance, the optimization of quantum algorithms will play a pivotal role in unlocking the full potential of quantum computing. Quantum algorithm designers and researchers collaborate to develop efficient, practical, and scalable quantum algorithms that address real-world optimization challenges. In the quest for quantum advantage, where quantum computers outperform classical computers on specific tasks, algorithm optimization is a crucial step towards realizing the full capabilities of quantum technology. In summary, quantum algorithm optimization is a multidisciplinary field that spans quantum mechanics, computer science, mathematics, and practical applications. It involves enhancing the efficiency and performance of quantum algorithms while considering hardware constraints, error correction, and practicality. Quantum algorithms have the potential to revolutionize industries and solve complex optimization problems that were previously intractable for classical computers. The ongoing research and development in quantum algorithm optimization hold the promise of transformative advancements in computation, offering new solutions to challenges across various domains. Quantum algorithms have gained significant attention for their potential to solve specific problems more efficiently than classical algorithms. These specialized quantum algorithms leverage the unique properties of quantum mechanics, such as superposition and entanglement, to provide computational advantages for particular tasks. One of the most well-known quantum algorithms is Shor's algorithm, which is designed to factor large integers exponentially faster than the best-known classical algorithms. The ability to factor large numbers efficiently has profound implications for cryptography, as many encryption schemes rely on the difficulty of factoring large numbers. Shor's algorithm poses a significant threat to classical encryption methods and has sparked interest in post-quantum cryptography as a response. Another groundbreaking quantum algorithm is Grover's algorithm, which accelerates the search for an item in an unsorted database from a classical $O(N)$ time complexity to a quantum $O(\sqrt{N})$ time complexity. This quadratic speedup has applications in database search, optimization, and solving various computational problems efficiently. Quantum algorithms are not limited to number theory and search problems; they also have applications in quantum chemistry. Quantum chemistry simulations are computationally expensive, but quantum algorithms, like the quantum phase estimation algorithm, promise to provide exponential speedup in solving problems related to molecular modeling and chemical reactions. These advancements could revolutionize drug discovery and materials science by enabling the accurate prediction of molecular properties and behavior. Quantum machine learning algorithms, such as quantum support vector machines and quantum neural networks, offer the potential to speed up complex data analysis tasks. They can efficiently handle large datasets and optimize machine learning models, making them valuable tools for data-driven industries and artificial intelligence research. Quantum algorithms are not confined to specific industries; they have broad applications in optimization problems. For example, the quantum approximate *optimization algorithm (QAOA) aims to find approximate solutions to combinatorial optimization problems, including the well-known traveling salesman problem. By harnessing quantum parallelism and interference, QAOA can explore solution spaces more efficiently than classical algorithms, offering significant improvements in solving complex optimization challenges. Quantum algorithms also extend their reach to graph theory and network analysis. The quantum PageRank algorithm, inspired by Google's PageRank, is designed to determine the importance of nodes in a large network. This algorithm has applications in various fields, from ranking web pages to identifying key nodes in complex networks like social networks and transportation systems. Quantum algorithms are continually evolving, and new quantum algorithms are being developed to address an array of practical problems. The quantum simulation of physical systems is a crucial area of research. Quantum computers can simulate quantum systems more efficiently than classical computers, allowing for the study of quantum materials, chemical reactions, and condensed matter physics. This capability could lead to the discovery of new materials with unique properties or the optimization of energy-efficient processes. Quantum algorithms also play a role in optimization problems with real-world applications, such as supply chain management. Efficiently optimizing supply chain routes, inventory management, and production scheduling can have significant economic and environmental impacts. Quantum algorithms offer the potential to address these complex optimization challenges effectively. The quantum approximate optimization algorithm (QAOA) and quantum annealing approaches can be adapted to tackle supply chain optimization tasks. In finance, quantum algorithms have the potential to revolutionize portfolio optimization and risk assessment. Optimizing investment portfolios to maximize returns while minimizing risk is a computationally intensive task. Quantum computing offers the possibility of rapidly exploring numerous portfolio combinations and identifying optimal investment strategies. Moreover, quantum algorithms can improve risk assessment by efficiently modeling complex financial systems and predicting market behavior. The potential impact of quantum algorithms on finance extends to options pricing, algorithmic trading, and fraud detection. Machine learning and artificial intelligence are domains where quantum algorithms can enhance model training, data analysis, and pattern recognition. Quantum machine learning algorithms, such as quantum support vector machines and quantum neural networks, offer the advantage of handling large datasets and optimizing complex models more efficiently. This can lead to advancements in natural language processing, image recognition, and recommendation systems. Quantum algorithms for machine learning also have applications in healthcare, where they can accelerate the analysis of medical data, drug discovery, and genomics research. The quantum advantage in machine learning tasks holds the potential to revolutionize personalized medicine and diagnostics. Quantum algorithms are not just limited to optimization and machine learning; they also impact the field of cryptography. Post-quantum cryptography is a critical area of research aimed at developing encryption methods that are secure against quantum attacks. Quantum algorithms like Shor's algorithm have the capability to factor large numbers exponentially faster than classical algorithms, which threatens the security of many encryption schemes. To address this vulnerability, researchers are working on developing quantum-resistant encryption techniques that can withstand quantum attacks. Quantum key distribution, which relies on the principles of quantum mechanics, offers a* secure method for exchanging encryption keys, immune to quantum attacks. Quantum-resistant cryptographic algorithms are essential to protect sensitive information in a post-quantum era. Quantum algorithms also have applications in quantum communications, where they enable secure and efficient transmission of quantum information. Quantum teleportation, for instance, allows the instantaneous transfer of quantum states between distant locations using entanglement. This concept has implications for secure communication and quantum networking, enabling the development of quantum internet protocols. In summary, quantum algorithms have the potential to revolutionize various fields by solving specific problems more efficiently than classical algorithms. From cryptography to finance, supply chain management, machine learning, and quantum chemistry, quantum algorithms offer a new paradigm for tackling complex computational challenges. As quantum hardware continues to advance, quantum algorithms will play a pivotal role in unlocking the full potential of quantum computing, ushering in a new era of technological advancement and scientific discovery.

Chapter 4: Quantum Machine Learning and Optimization

Quantum enhancements have the potential to revolutionize the field of machine learning by accelerating various aspects of the machine learning pipeline. Machine learning encompasses a wide range of techniques that enable computers to learn from data and make predictions or decisions without being explicitly programmed. Quantum computing, with its unique properties like superposition and entanglement, offers the promise of accelerating classical machine learning algorithms. One of the key advantages of quantum computing for machine learning is its ability to process and analyze vast amounts of data more efficiently. Quantum computers can represent and manipulate data in superposition, allowing them to explore multiple possibilities simultaneously. This parallelism can lead to significant speedups in data preprocessing, feature selection, and data augmentation, all essential steps in the machine learning workflow. Quantum machine learning algorithms are designed to harness the power of quantum parallelism to perform computations more efficiently. Quantum support vector machines (QSVMs) are one such algorithm that can provide advantages in classification tasks. QSVMs use quantum features and quantum interference to find optimal hyperplanes for classifying data points in high-dimensional feature spaces. Quantum machine learning also includes quantum versions of classical algorithms, such as quantum k-means clustering and quantum principal component analysis (PCA). These quantum counterparts aim to provide speedup and efficiency improvements over classical methods. Quantum computing can also expedite the training of machine learning models. Quantum-enhanced optimization algorithms, such as the quantum approximate optimization algorithm (QAOA), can be employed to find optimal model parameters more quickly. This is particularly valuable in deep learning, where training large neural networks can be computationally intensive. Quantum-enhanced optimization techniques can accelerate convergence and reduce the time required to train complex models. Quantum machine learning algorithms are not limited to classical data analysis; they can also be applied to quantum data. Quantum information processing involves quantum states and operations, making it natural to use quantum computing for quantum data analysis. Quantum algorithms can help discover patterns, correlations, and insights within quantum datasets, which are prevalent in quantum physics experiments and quantum technologies. Quantum-enhanced feature selection is another area where quantum computing can benefit machine learning. Feature selection involves choosing the most relevant features or variables from a dataset while discarding irrelevant or redundant ones. Quantum algorithms can explore a large feature space more efficiently than classical algorithms, leading to better feature selection and improved model performance. Quantum computing can also address challenges in optimization problems commonly encountered in machine learning. Many machine learning tasks involve finding optimal solutions to complex objective functions, such as minimizing loss functions in neural network training or optimizing hyperparameters. Quantum-enhanced optimization algorithms, like the quantum approximate optimization algorithm (QAOA) and quantum annealing, can expedite the search for optimal solutions. Quantum annealing, in particular, leverages quantum effects to explore solution spaces more efficiently and find global optima for challenging optimization problems. Quantum computing can *contribute to federated learning, a decentralized approach to machine learning where models are trained across multiple devices or servers while keeping data localized. Privacy concerns and data security are crucial in federated learning, and quantum cryptography can provide secure communication and data sharing among participants. Quantum key distribution (QKD) protocols enable secure and unforgeable communication channels, ensuring that federated learning participants can exchange model updates and gradients without compromising data privacy. Quantum computing also holds promise in addressing challenges related to explainability and interpretability in machine learning. Quantum machine learning models may offer new ways to extract and represent features from data, potentially leading to more interpretable model outputs. This can be valuable in applications where understanding the model's decision-making process is critical, such as healthcare or autonomous vehicles. Quantum machine learning algorithms are still in the early stages of development, and their practical implementation faces several challenges. One significant challenge is the limited availability of quantum hardware with the qubits and gate operations needed for quantum machine learning algorithms. Quantum computers are still in their infancy, and large-scale, fault-tolerant quantum processors are yet to become widely accessible. As a result, quantum machine learning is often conducted using quantum simulators or noisy intermediate-scale quantum (NISQ) devices. These devices have limitations in terms of qubit count, gate fidelity, and coherence times, which can affect the scalability and performance of quantum machine learning algorithms. Error correction is another critical challenge in quantum machine learning. Quantum computers are susceptible to errors caused by decoherence, gate imperfections, and environmental factors. To mitigate these errors, researchers are working on developing quantum error correction codes and fault-tolerant quantum computing techniques. These advances are essential for realizing the full potential of quantum-enhanced machine learning. Quantum machine learning also requires specialized quantum software development and quantum programming skills. Developing quantum algorithms and adapting classical machine learning algorithms to quantum hardware demands expertise in quantum information theory, quantum gate operations, and quantum circuit design. Quantum machine learning researchers often collaborate across fields, combining expertise in quantum physics, computer science, and machine learning. The integration of quantum machine learning into existing machine learning workflows and toolkits is an ongoing area of research. Quantum software platforms, quantum programming languages, and quantum machine learning libraries are emerging to facilitate the development and implementation of quantum-enhanced machine learning algorithms. Quantum machine learning holds the potential to accelerate advancements in various fields, including healthcare, finance, materials science, and artificial intelligence. Applications range from drug discovery and optimization in healthcare to portfolio optimization and risk assessment in finance. Quantum-enhanced machine learning algorithms have the potential to solve complex problems more efficiently, ultimately driving innovation and scientific discovery. As quantum hardware continues to advance, the field of quantum machine learning will likely expand, providing new tools and techniques for solving challenging real-world problems. In summary, quantum enhancements offer significant potential for accelerating various aspects of the machine learning pipeline. Quantum algorithms and quantum computing techniques can expedite data preprocessing, feature selection, optimization, and model training. Quantum machine learning also extends to quantum data analysis and quantum information processing. However, the field faces challenges related to the availability of quantum hardware, error correction, and the need for specialized quantum programming skills. Despite these challenges, quantum machine learning holds promise in revolutionizing fields and industries by providing efficient solutions to complex problems. Quantum optimization techniques are at the forefront of quantum computing, offering the potential to solve complex optimization problems more efficiently than classical methods. These techniques harness the unique properties of quantum mechanics, such as superposition and entanglement, to explore solution spaces and find optimal solutions. One of the key advantages of quantum optimization is its ability to handle large-scale combinatorial and continuous optimization problems. Classical optimization algorithms often face exponential time complexity when dealing with such problems, making them infeasible for practical use. Quantum computers, on the other hand, have the potential to provide exponential speedup for certain optimization tasks. Quantum optimization techniques are not limited to a single algorithm or approach; they encompass a variety of methods and strategies. Quantum annealing, for example, is a quantum optimization technique inspired by the annealing process in metallurgy. In quantum annealing, a quantum system is prepared in a particular initial state, and its energy is gradually lowered to reach the ground state, which corresponds to the optimal solution of the optimization problem. Quantum annealers, such as those developed by D-Wave Systems, leverage quantum effects to explore solution spaces efficiently. Quantum approximate optimization algorithms (QAOAs) are another class of quantum optimization techniques. QAOAs use a sequence of quantum gates to evolve a quantum state that represents a candidate solution. By optimizing the gate parameters, QAOAs aim to find the optimal solution by exploiting quantum parallelism and interference effects. Quantum-enhanced optimization algorithms are not limited to combinatorial problems; they can also address continuous optimization tasks. For instance, quantum algorithms like the quantum approximate Bayesian computation (qABC) algorithm aim to sample from complex probability distributions efficiently. This capability is valuable in Bayesian optimization, machine learning, and statistical analysis. Quantum optimization techniques offer advantages in various domains, including logistics, finance, materials science, and machine learning. In logistics and supply chain management, quantum optimization can optimize routes, schedules, and inventory management to reduce costs and improve efficiency. Optimizing complex supply chain networks with classical methods can be computationally challenging, but quantum algorithms can handle the complexity more efficiently. Financial optimization tasks, such as portfolio management and risk assessment, can benefit from quantum computing's ability to explore large solution spaces rapidly. Efficiently managing portfolios with thousands of assets or optimizing trading strategies in real-time are examples of challenges where quantum optimization techniques may provide a competitive advantage. Materials science and drug discovery also stand to gain from quantum optimization. Quantum computers can simulate the behavior of molecules and materials more accurately than classical computers, facilitating the discovery of new materials with desired properties. Quantum optimization algorithms can help researchers explore the vast space of possible molecular structures efficiently.*

Quantum machine learning models and optimization techniques are increasingly integrated into the machine learning workflow. Quantum support vector machines (QSVMs), for example, offer advantages in classification tasks by finding optimal hyperplanes in high-dimensional feature spaces. Quantum-enhanced optimization algorithms can speed up model training and hyperparameter optimization in deep learning. Despite their potential, quantum optimization techniques face several challenges. One significant challenge is the availability and scalability of quantum hardware. Currently, quantum computers with a sufficient number of qubits and gate operations for practical optimization tasks are limited. Moreover, the qubits in current quantum devices are susceptible to errors due to decoherence and gate imperfections. Quantum error correction techniques are still in development and add an overhead that can limit the performance of quantum optimization algorithms. Another challenge is the need for quantum programming expertise. Developing and implementing quantum optimization algorithms require knowledge of quantum gate operations, quantum circuits, and quantum programming languages. The field of quantum computing is rapidly evolving, and researchers are working on improving the scalability, error correction, and usability of quantum hardware. Hybrid quantum-classical approaches are also explored, where classical computers assist in solving optimization problems alongside quantum computers. These hybrid approaches aim to leverage the strengths of both classical and quantum computing to tackle complex optimization tasks efficiently. Quantum optimization techniques are not confined to academia and research; they are gaining attention from industries and organizations looking for solutions to challenging optimization problems. Companies are exploring quantum computing for applications in logistics, finance, and materials science. Startups and technology companies are developing quantum software and tools to harness the power of quantum optimization for practical purposes. Quantum-enhanced optimization is a dynamic and interdisciplinary field that combines quantum physics, computer science, mathematics, and practical applications. It holds the promise of transforming industries and revolutionizing the way complex optimization problems are solved. In summary, quantum optimization techniques are poised to revolutionize the field of optimization by offering exponential speedup for solving complex problems. Quantum annealing, quantum approximate optimization algorithms (QAOAs), and quantum-enhanced optimization methods provide valuable tools for tackling large-scale optimization challenges. Despite challenges related to quantum hardware and programming, quantum optimization is making strides in various domains, from logistics and finance to materials science and machine learning. As quantum technology continues to advance, quantum optimization will likely play a pivotal role in addressing complex real-world problems and driving innovation across industries.

Chapter 5: Quantum Simulation and Quantum Chemistry

Advanced quantum simulations represent a cutting-edge and promising application of quantum computing technology. These simulations aim to model and understand complex quantum systems, which can be challenging or even impossible for classical computers to handle efficiently. Quantum simulations leverage the inherent quantum properties of superposition and entanglement to explore quantum systems' behavior and properties. One of the primary motivations for advanced quantum simulations is to study and understand quantum materials. Quantum materials exhibit unique properties, such as high-temperature superconductivity and topological insulator behavior, that hold immense potential for technological advancements. However, predicting and engineering these materials require a deep understanding of quantum interactions, which can be achieved through advanced quantum simulations. Quantum simulations have already been instrumental in studying and discovering new materials with exceptional properties. For example, advanced quantum simulations have played a crucial role in identifying potential high-temperature superconductors, which could revolutionize energy transmission and storage technologies. Quantum simulations also contribute to the development of new materials for efficient batteries, catalysts, and quantum technologies. Quantum simulations extend beyond materials science into other areas of physics, such as quantum chemistry. Quantum chemistry aims to understand molecular interactions and predict molecular properties accurately. Quantum simulations provide a powerful tool for solving the Schrödinger equation, which describes the quantum behavior of molecules. These simulations enable researchers to explore chemical reactions, optimize catalysts, and design new drugs with higher precision. Quantum computers have the potential to revolutionize drug discovery by simulating complex biological systems and predicting drug interactions accurately. Another significant application of advanced quantum simulations is quantum field theory. Quantum field theory is a fundamental framework in theoretical physics used to describe the behavior of elementary particles and their interactions. Simulating quantum field theories accurately is challenging due to their intricate mathematical structures and high-dimensional Hilbert spaces. Quantum simulations offer a path to explore and test quantum field theories under various conditions, potentially leading to new insights into the fundamental laws of the universe. Quantum simulations can also address problems in condensed matter physics, where understanding the behavior of many interacting particles is essential. For instance, simulating the behavior of electrons in a solid-state system can shed light on phenomena like magnetism, topological phases, and quantum phase transitions. Advanced quantum simulations have the potential to uncover new quantum materials with unique electronic properties. These materials could have applications in quantum computing, spintronics, and advanced electronics. One of the most celebrated quantum algorithms for advanced simulations is the quantum phase estimation algorithm. This algorithm can efficiently estimate the eigenvalues of a unitary operator, making it valuable for simulating quantum systems and solving quantum chemistry problems. Quantum phase estimation leverages the principles of quantum parallelism and interference to provide exponential speedup compared to classical methods. Variational quantum algorithms are another category of algorithms widely used in advanced quantum simulations.

Variational quantum algorithms employ parameterized quantum circuits to approximate the ground state of a quantum system. These algorithms are versatile and can adapt to various quantum systems, making them valuable tools for quantum simulations. Incorporating noise and error mitigation techniques is crucial in advanced quantum simulations. Quantum hardware is susceptible to noise, decoherence, and gate imperfections, which can affect the accuracy of simulations. Quantum error correction codes and error mitigation strategies aim to minimize the impact of these errors, ensuring reliable simulation results. Hybrid quantum-classical approaches are gaining prominence in advanced quantum simulations. These approaches combine classical and quantum resources to solve complex problems efficiently. Classical computers are used for pre- and post-processing tasks, while quantum computers handle the core of the simulation. Hybrid simulations can overcome the limitations of current quantum hardware and improve the scalability of advanced quantum simulations. Quantum simulations also benefit from advances in quantum hardware. Quantum processors with more qubits, longer coherence times, and lower error rates are essential for simulating larger and more complex quantum systems accurately. Companies and research institutions are actively working on developing and scaling quantum hardware for advanced simulations. Quantum cloud platforms and quantum-as-a-service offerings provide researchers and scientists with access to quantum computing resources for simulations and experiments. These platforms aim to democratize quantum technology and accelerate progress in quantum simulations. Challenges in advanced quantum simulations include optimizing quantum circuits for specific problems, mitigating noise and errors, and developing efficient quantum algorithms. Researchers are continuously working on improving the accuracy, scalability, and performance of quantum simulations. Advanced quantum simulations hold immense promise in solving complex problems in materials science, quantum chemistry, condensed matter physics, and quantum field theory. These simulations enable researchers to explore new materials, understand fundamental physical phenomena, and design novel technologies. As quantum hardware matures and quantum algorithms evolve, advanced quantum simulations will continue to push the boundaries of scientific discovery and technological innovation.

Quantum chemistry applications represent a profound intersection of quantum physics and chemistry, offering insights into the behavior of atoms and molecules at the quantum level. These applications have wide-reaching implications for fields such as materials science, drug discovery, and understanding the fundamental processes that govern chemical reactions. Quantum chemistry involves the use of quantum mechanical principles and computational methods to predict and understand the electronic structure and properties of molecules. One of the central tasks in quantum chemistry is solving the Schrödinger equation for a given molecular system. This equation describes the quantum state of electrons and nuclei in the molecule, including their energy levels and spatial distributions. Solving the Schrödinger equation provides crucial information about a molecule's electronic structure and its behavior in chemical reactions. Quantum chemistry applications can simulate chemical reactions, helping chemists and researchers understand reaction mechanisms and predict the products of reactions. This capability is invaluable in designing new chemical processes, optimizing reaction conditions, and discovering novel catalysts. Quantum chemistry also plays a vital role in the field of materials science. By simulating the electronic properties and behavior of materials at the quantum level, researchers can identify materials with desirable characteristics for specific applications. For instance, quantum chemistry applications have led to the discovery of materials with exceptional conductivity, magnetism, or superconductivity, which have implications for electronics and energy storage technologies. The study of quantum chemistry is instrumental in drug discovery and pharmaceutical research. Quantum simulations can predict the interactions between drug molecules and biological targets, helping researchers design more effective drugs with fewer side effects. Quantum chemistry applications also aid in understanding the behavior of biomolecules, such as proteins and DNA, at the molecular level, shedding light on the fundamental processes of life. Quantum chemistry calculations involve solving the electronic Schrödinger equation, which requires approximations due to the complexity of molecular systems. One widely used approximation method is density functional theory (DFT), which simplifies the many-body electronic problem by considering the electron density rather than individual electron wavefunctions. DFT has become a workhorse of quantum chemistry, enabling the study of large molecular systems and materials. Another essential quantum chemistry method is Hartree-Fock theory, which provides a self-consistent field approximation to the electronic structure. Hartree-Fock calculations serve as a foundation for more advanced electronic structure methods. Post-Hartree-Fock methods, such as configuration interaction (CI) and coupled cluster (CC) theory, account for electron correlation effects that are not fully captured by Hartree-Fock calculations. These methods are crucial for accurately predicting the properties of molecules with strongly correlated electrons, such as those involved in transition metal complexes and chemical reactions. Quantum chemistry applications often employ basis sets, which are sets of functions used to approximate electron wavefunctions. The choice of basis set can significantly impact the accuracy of quantum chemistry calculations. Advanced basis sets, such as Gaussian basis sets, offer greater flexibility and accuracy in representing molecular wavefunctions. Quantum chemistry calculations can be computationally demanding, especially for large molecular systems. As a result, quantum chemists rely on high-performance computing clusters and supercomputers to carry out complex simulations. Quantum chemistry software packages, such as Gaussian, GAMESS, and NWChem, provide a range of methods and tools for performing electronic structure calculations. These software packages enable researchers to tackle a wide variety of chemical problems, from predicting molecular properties to simulating chemical reactions. Quantum chemistry applications extend beyond ground-state calculations to include excited-state properties, which are essential for understanding molecular spectroscopy and photochemical processes. Time-dependent density functional theory (TDDFT) and configuration interaction with singles and doubles (CISD) are examples of methods used to study excited states. Quantum chemistry is also employed in the field of computational spectroscopy, where it aids in interpreting experimental data and predicting the spectra of molecules. Nuclear magnetic resonance (NMR) spectroscopy and infrared (IR) spectroscopy, among others, benefit from quantum chemistry simulations. Quantum chemistry applications have had a profound impact on our understanding of chemical reactions and molecular properties. They have been instrumental in predicting the behavior of molecules in diverse environments, including those relevant to *atmospheric chemistry and astrochemistry. In atmospheric chemistry, quantum chemistry calculations help researchers understand the formation and transformation of molecules in the Earth's atmosphere, leading to insights into air quality and climate change. Astrochemistry explores the chemistry of molecules in space, from the interstellar medium to the atmospheres of distant planets. Quantum chemistry simulations provide valuable data for interpreting astronomical observations and unraveling the chemistry of the cosmos. Quantum chemistry applications continue to evolve with advancements in both quantum algorithms and quantum hardware. Quantum computers have the potential to solve electronic structure problems with unprecedented efficiency, offering new avenues for tackling complex chemical systems. Quantum computing companies and research institutions are actively developing quantum algorithms and hardware optimized for quantum chemistry applications. Quantum chemistry is a dynamic field that bridges fundamental science with practical applications. It empowers researchers to design new materials, develop innovative drugs, understand chemical reactions, and explore the mysteries of the universe. As quantum technology advances, quantum chemistry applications will continue to drive scientific discovery and technological innovation.*

Chapter 6: Error Correction and Fault-Tolerant Quantum Computing

Advanced error correction schemes play a critical role in the development and reliability of quantum computers. Quantum computers, while promising, are inherently susceptible to errors due to decoherence, gate imperfections, and environmental factors. These errors can disrupt the correct execution of quantum algorithms, limiting the practicality of quantum computing. Error correction schemes are essential to mitigate these issues and enable fault-tolerant quantum computing. One of the fundamental principles of quantum error correction is encoding quantum information redundantly across multiple qubits. This redundancy allows the correction of errors without destroying the quantum information, akin to classical error-correcting codes. The most well-known quantum error correction code is the surface code, which encodes logical qubits by arranging physical qubits in a two-dimensional lattice. In the surface code, qubits are measured in a specific pattern, and the outcomes are used to identify and correct errors. Surface code-based error correction can achieve extremely low error rates when implemented fault-tolerantly. However, it requires a large number of physical qubits for each logical qubit, making it resource-intensive. Advanced error correction schemes aim to reduce the resource overhead while maintaining or even improving error correction capabilities. One approach to more efficient error correction is the use of concatenated codes, where multiple layers of encoding are applied successively. These codes can achieve a high threshold error rate while requiring fewer physical qubits than the surface code. Concatenated codes offer a compromise between error correction capability and resource efficiency. Another innovative approach to error correction is topological codes, such as the toric code and the color code. These codes leverage the topology of a surface to encode quantum information. Errors manifest as topological defects, making them easier to detect and correct. Topological codes have the potential to provide highly fault-tolerant quantum computation with reduced overhead. Advanced error correction schemes also consider the use of non-binary codes, such as qudit codes, which encode information in higher-dimensional quantum systems. Qudits can carry more information than qubits, potentially allowing for more efficient error correction. Non-binary codes are an area of active research in quantum error correction. Bosonic codes, which deal with quantum information carried by continuous variables like photons, offer an alternative approach to error correction. These codes exploit the properties of continuous-variable systems to protect quantum information from errors. Bosonic codes are particularly relevant for quantum communication and quantum optics experiments. Quantum error correction is not limited to specific quantum hardware; it encompasses a range of architectures, including superconducting qubits, trapped ions, and photonic qubits. Each platform has its unique error characteristics and challenges, necessitating tailored error correction schemes. Superconducting qubits, for example, are susceptible to both coherent and incoherent errors, making it crucial to address both types in error correction strategies. Trapped ion quantum computers face challenges related to qubit connectivity and gate fidelities, which impact error correction approaches. Photonic quantum computers rely on encoding quantum information in the properties of photons and face challenges such as photon loss and detector inefficiencies. Advanced error correction schemes aim to address these platform-specific issues while enhancing the overall fault tolerance of quantum computers. Error correction also extends to quantum networking, where quantum information is transmitted over long distances through quantum communication channels. Quantum key distribution (QKD) protocols, such as the BBM92 and E91 protocols, provide secure communication channels based on the principles of quantum mechanics. Error correction and privacy amplification techniques are essential components of QKD to ensure secure key distribution. These techniques correct errors that may arise during the transmission of quantum states and distill a secret key shared between two parties. Quantum error correction is intertwined with quantum cryptography, as secure quantum communication relies on error-free transmission of quantum states. The development of fault-tolerant quantum computing requires both hardware advancements and sophisticated error correction techniques. Quantum error correction codes, such as the surface code and concatenated codes, are designed to detect and correct errors that occur during quantum computation. However, implementing these codes in practice is challenging due to the resource overhead and sensitivity to errors during error correction itself. Fault-tolerant quantum computing aims to build a quantum computer that can execute quantum algorithms reliably, even in the presence of errors. One of the key components of fault-tolerant quantum computing is fault-tolerant quantum gates. These gates are designed to perform quantum operations accurately despite the presence of errors in the underlying qubits. Threshold theorems provide mathematical proofs that fault-tolerant quantum computing is possible if the error rate of individual qubits and gates is below a certain threshold. The threshold theorem establishes a clear goal for quantum error correction efforts:

to reduce error rates below the threshold. To achieve this goal, researchers explore various quantum error mitigation techniques, including quantum error correction codes, gate-level error correction, and error-avoiding quantum algorithms. Quantum error correction codes, such as the surface code and concatenated codes, encode quantum information in redundant patterns that allow for the detection and correction of errors. These codes rely on syndrome measurements to identify errors and then apply corrective operations to recover the quantum state. Gate-level error correction techniques aim to make quantum gates more robust against errors by implementing error-correcting codes directly within gate operations. Quantum algorithms, such as the Variational Quantum Eigensolver (VQE) and Quantum Approximate Optimization Algorithm (QAOA), are designed to be less sensitive to noise and errors. These algorithms are used in conjunction with error mitigation strategies to achieve accurate results on noisy quantum hardware. Quantum error mitigation techniques also include error extrapolation, which estimates the noise level and adjusts algorithm parameters accordingly to reduce errors. Additionally, researchers are exploring machine learning approaches to error mitigation, where neural networks are trained to correct errors in quantum data. Quantum error correction is an ongoing and interdisciplinary field of research that combines quantum physics, computer science, and information theory. Efforts are directed toward achieving fault-tolerant quantum computing and improving the reliability of quantum technologies. Quantum hardware and software advancements are crucial for making quantum computers and quantum communication systems more practical and robust. Quantum error correction is a cornerstone of these efforts, aiming to overcome the inherent challenges posed by quantum noise and errors. As quantum technology continues to evolve, the field of quantum error correction will play a pivotal role in realizing the full potential of quantum computing and quantum communication.

Achieving fault tolerance in quantum systems is a central challenge in the development of practical and reliable quantum technologies. Quantum systems are inherently fragile, as they are susceptible to errors caused by decoherence, gate imperfections, and environmental interactions. Fault tolerance is the ability to perform quantum computations accurately, even in the presence of errors. Without fault tolerance, the potential power of quantum computing and quantum communication is limited. The threshold theorem provides a rigorous foundation for achieving fault tolerance in quantum systems. This theorem sets a threshold error rate for individual qubits and quantum gates, below which fault-tolerant quantum computation becomes feasible. To meet this threshold, quantum error correction codes, advanced gate-level error correction techniques, and error-avoiding quantum algorithms are essential. Quantum error correction codes are at the heart of fault-tolerant quantum computing. These codes encode quantum information redundantly in a way that allows for the detection and correction of errors. The most well-known quantum error correction code is the surface code, which organizes qubits in a two-dimensional lattice and relies on syndrome measurements to identify errors and correct them. The surface code, while powerful, demands a large number of physical qubits for each logical qubit, making it resource-intensive. Advanced error correction schemes aim to reduce this resource overhead while maintaining or improving error correction capabilities. One approach is the use of concatenated codes, which apply multiple layers of encoding successively. These codes offer a balance between error *correction capability and resource efficiency. Another innovative approach is the utilization of topological codes like the toric code and color code. These codes exploit the topology of a surface to encode quantum information, with errors manifesting as topological defects that are easier to detect and correct. Advanced error correction schemes also explore non-binary codes, such as qudit codes, which encode information in higher-dimensional quantum systems. Qudits offer the potential to carry more information than qubits, making error correction more efficient. Non-binary codes are an active area of research in quantum error correction. Bosonic codes, designed for quantum systems with continuous variables, represent another avenue for error correction. These codes leverage the properties of continuous-variable systems to protect quantum information from errors and are relevant for quantum communication and quantum optics. Error correction efforts are not limited to specific quantum hardware; they encompass various platforms, including superconducting qubits, trapped ions, and photonic qubits. Each platform has its unique error characteristics and challenges, necessitating tailored error correction strategies. Superconducting qubits, for instance, are susceptible to both coherent and incoherent errors, requiring mitigation techniques for both types. Trapped ion quantum computers face challenges related to qubit connectivity and gate fidelities, impacting error correction approaches. Photonic quantum computers rely on encoding quantum information in the properties of photons and face challenges like photon loss and detector inefficiencies. To achieve fault tolerance, researchers investigate a range of quantum error mitigation techniques, including quantum error correction codes, gate-level error correction, and error-avoiding quantum algorithms. Quantum error correction codes, such as the surface code and concatenated codes, play a vital role in detecting and correcting errors. These codes rely on syndrome measurements to identify errors and apply corrective operations to recover the quantum state. Gate-level error correction techniques aim to make quantum gates more robust against errors by implementing error-correcting codes directly within gate operations. Quantum algorithms, such as the Variational Quantum Eigensolver (VQE) and Quantum Approximate Optimization Algorithm (QAOA), are designed to be less sensitive to noise and errors. These algorithms, when combined with error mitigation strategies, help achieve accurate results on noisy quantum hardware. Quantum error mitigation techniques also encompass error extrapolation, which estimates the noise level and adjusts algorithm parameters to reduce errors. Additionally, machine learning approaches are explored for error mitigation, where neural networks are trained to correct errors in quantum data. Quantum error correction is an interdisciplinary field that combines quantum physics, computer science, and information theory. Efforts in quantum error correction are directed toward achieving fault-tolerant quantum computing and improving the reliability of quantum technologies. Quantum hardware and software advancements are crucial for making quantum computers and quantum communication systems more practical and robust. Quantum error correction represents a cornerstone of these efforts, addressing the fundamental challenges posed by quantum noise and errors. As quantum technology continues to evolve, the field of quantum error correction will play a pivotal role in realizing the full potential of quantum computing and quantum communication.*

Chapter 7: Quantum Hardware and Quantum Processors

Cutting-edge quantum hardware stands at the forefront of scientific and technological innovation, promising to reshape the landscape of computing and communication. Quantum hardware encompasses a diverse range of physical platforms, each with its unique strengths and challenges. Among these platforms are superconducting qubits, trapped ions, photonic qubits, and topological qubits. Superconducting qubits are a leading contender in the race to build practical quantum computers. These qubits leverage the principles of superconductivity to achieve low-noise and high-fidelity quantum operations. They are characterized by their compatibility with existing semiconductor fabrication techniques, allowing for the integration of qubits with classical components on the same chip. Superconducting qubits have achieved remarkable progress in recent years, with devices containing tens to hundreds of qubits now available. However, challenges remain in improving qubit coherence times and reducing error rates. Trapped ions offer another promising platform for quantum computing. In this approach, ions are trapped and manipulated using electromagnetic fields. Trapped ion qubits are known for their long coherence times and low error rates. They have demonstrated exquisite control over individual qubits and the ability to entangle multiple qubits. Trapped ion quantum computers have already achieved quantum advantage in specific applications. However, scaling up the number of qubits and minimizing crosstalk between qubits are ongoing challenges. Photonic qubits, which use photons as quantum bits, excel in quantum communication and quantum key distribution. Photons are excellent carriers of quantum information, as they can travel long distances with minimal decoherence. Photonic quantum technologies have enabled secure quantum communication protocols and quantum cryptography. Moreover, photonic quantum computers hold the potential to perform quantum simulations efficiently. Topological qubits, inspired by topological quantum states of matter, offer a unique approach to error-resistant quantum computation. These qubits rely on non-local properties of quantum states to protect against local errors. Topological qubits are intrinsically fault-tolerant, making them an attractive candidate for large-scale quantum computers. However, they are still in the early stages of development, with significant technical challenges to overcome. One of the fundamental challenges in quantum hardware is achieving and maintaining qubit coherence. Quantum coherence refers to the ability of a qubit to maintain a superposition of states without decohering into a classical state. Decoherence arises from interactions with the environment, thermal fluctuations, and imperfections in control electronics. Efforts to extend qubit coherence times involve cryogenic cooling, error correction codes, and advanced pulse sequences. Moreover, quantum error correction techniques, such as the surface code, are designed to protect quantum information from errors and extend the fault tolerance threshold. Quantum error correction is an essential component of quantum hardware development, aiming to reduce the impact of noise and errors on quantum computations. Scaling up the number of qubits is another significant challenge in quantum hardware. Building large-scale quantum computers requires addressing issues related to qubit connectivity, gate fidelity, and resource overhead. Entangling qubits that are not directly connected poses a challenge in many quantum hardware platforms. Developing efficient quantum gates that operate with high fidelity is crucial for scaling up quantum devices. Furthermore, mitigating the resource overhead of error correction codes, which often require many physical qubits to protect a single logical qubit, is essential for practical quantum computing. Quantum hardware development is a multidisciplinary endeavor that involves expertise in physics, materials science, engineering, and computer science. Researchers and engineers work collaboratively to design, fabricate, and operate quantum devices. In recent years, quantum computing companies and startups have emerged as key players in the quantum hardware ecosystem. These companies are driving advancements in quantum hardware and making quantum computing resources more accessible through cloud platforms and quantum-as-a-service offerings. Major technology companies, such as IBM, Google, and Intel, are heavily invested in quantum hardware research and development, striving to achieve quantum supremacy and explore the practical applications of quantum computers. Academic institutions and research laboratories around the world are also making significant contributions to the field of quantum hardware. International collaborations and open-source initiatives facilitate knowledge sharing and accelerate progress in quantum hardware development. The development of cutting-edge quantum hardware is not limited to quantum computers; it extends to quantum sensors, quantum communication devices, and quantum simulators. Quantum sensors, such as atomic clocks and magnetometers, leverage quantum phenomena to achieve unparalleled precision in measurements. These sensors have applications in navigation, geophysics, and fundamental physics experiments. Quantum communication devices enable secure and unbreakable communication channels based on the principles of quantum key distribution. Quantum simulators, which use quantum hardware to mimic the behavior of complex quantum systems, offer insights into quantum materials, chemistry, and fundamental physics. Quantum hardware represents a pivotal point in the journey towards practical quantum technologies. As quantum devices continue to advance, researchers and engineers are pushing the boundaries of what is possible in quantum computing, communication, and sensing. The road ahead is filled with both challenges and opportunities, but the potential rewards are groundbreaking, promising to revolutionize industries, solve complex problems, and uncover the mysteries of the quantum world. Quantum processor architectures represent the core of quantum computing systems, shaping the way quantum information is processed and manipulated. These architectures are fundamental to harnessing the power of quantum mechanics for computation and solving complex problems. At the heart of a quantum processor lies the qubit, the quantum counterpart of the classical bit. Unlike classical bits that can be either 0 or 1, qubits can exist in a superposition of states, representing both 0 and 1 simultaneously. This property enables quantum processors to explore multiple computational paths in parallel, providing the potential for exponential speedup in specific tasks. Quantum processors are designed with the goal of implementing quantum gates, which are the quantum analogs of classical logic gates. Quantum gates manipulate qubits to perform various quantum operations, including entanglement, superposition, and quantum parallelism. There are several quantum processor architectures, each with its unique approach to realizing and controlling qubits. One of the most widely used quantum processor architectures is based on superconducting qubits. Superconducting qubits are tiny circuits made from superconducting materials that can carry electrical currents *without resistance. These qubits are operated at extremely low temperatures, typically close to absolute zero, to exploit their quantum properties. Superconducting qubits are known for their scalability and compatibility with semiconductor fabrication techniques. Major quantum computing companies, such as IBM, Google, and Rigetti, have developed superconducting qubit-based processors. Trapped ion quantum processors represent another prominent architecture. In this approach, ions are trapped and manipulated using electromagnetic fields. Trapped ion qubits are prized for their long coherence times and high gate fidelities. They enable exquisite control over individual qubits and the creation of entangled states. Leading companies like IonQ and Honeywell have developed trapped ion quantum processors. Photonic quantum processors use photons as qubits, leveraging the properties of light to perform quantum computations. Photonic qubits are excellent carriers of quantum information, as they can travel long distances with minimal decoherence. Photonic processors are particularly well-suited for quantum communication and quantum key distribution. Topological quantum processors are an emerging architecture inspired by topological quantum states of matter. These processors rely on non-local properties of quantum states to protect against errors. Topological qubits are considered intrinsically fault-tolerant and have the potential for large-scale quantum computing. However, they are still in the early stages of development. Quantum processor architectures also include hybrid approaches that combine multiple qubit types. For instance, combining superconducting qubits with photonic interconnects has the potential to improve qubit connectivity and information transfer between different qubit types. Another key consideration in quantum processor architectures is qubit connectivity. Connecting qubits is essential for implementing quantum gates and entangling qubits effectively. The connectivity of a quantum processor can significantly impact the types of algorithms it can execute efficiently. Quantum hardware designers work to optimize qubit connectivity while minimizing crosstalk and unwanted interactions between qubits. Gate fidelities, which measure the accuracy of quantum gates, are critical in quantum processor architectures. High-fidelity gates are essential for performing quantum operations accurately and achieving reliable results. Efforts are ongoing to improve gate fidelities and reduce gate errors in quantum processors. Quantum processor architectures must also address the challenge of qubit coherence times. Quantum coherence refers to the ability of qubits to maintain quantum superpositions over time. Prolonged coherence times are crucial for executing complex quantum algorithms and minimizing error rates. Cryogenic cooling and error correction techniques are commonly employed to extend qubit coherence times. Resource overhead is another consideration in quantum processor architectures. Error correction codes, which are vital for achieving fault-tolerant quantum computing, often require a substantial number of physical qubits to protect a single logical qubit. Minimizing this resource overhead is essential for practical quantum computing. Quantum processor architectures are not limited to quantum computers; they extend to quantum simulators, quantum sensors, and quantum communication devices. Quantum simulators use quantum hardware to mimic the behavior of complex quantum systems, enabling researchers to study quantum materials, chemistry, and fundamental physics. Quantum sensors leverage quantum phenomena to achieve unparalleled precision in measurements, with applications in navigation, geophysics, and fundamental physics experiments. Quantum communication devices provide secure communication channels based on the principles of quantum key distribution, offering unbreakable encryption for sensitive information. Quantum processor architectures represent a dynamic and interdisciplinary field that combines expertise in physics, materials science, engineering, and computer science. Quantum hardware researchers and engineers work collaboratively to design, fabricate, and operate quantum devices. Quantum computing companies and startups play a crucial role in advancing quantum processor architectures and making quantum computing resources more accessible. Technology giants like IBM, Google, and Microsoft are heavily invested in quantum hardware research, striving to achieve quantum supremacy and explore the practical applications of quantum computers. Academic institutions and research laboratories worldwide contribute to the development of quantum processor architectures, fostering international collaborations and open-source initiatives. Quantum processor architectures are pivotal in unlocking the potential of quantum technologies, promising to revolutionize industries, solve complex problems, and deepen our understanding of the quantum world.*

Chapter 8: Quantum Programming Languages and Tools

Advanced quantum programming languages represent the foundation of quantum computing, serving as the interface between quantum hardware and quantum algorithms. These languages provide the tools and abstractions necessary to harness the power of quantum mechanics for solving complex problems. Quantum programming languages are designed to express quantum algorithms concisely and effectively, enabling quantum software developers to focus on the logic of their algorithms rather than the intricacies of quantum hardware. One of the most well-known quantum programming languages is Qiskit, developed by IBM. Qiskit offers a comprehensive suite of tools for quantum software development, including a high-level programming language, quantum simulators, and access to IBM's quantum hardware. Qiskit's programming language is Python-based, making it accessible to a broad community of developers. The language provides a user-friendly syntax for defining quantum circuits, composing quantum algorithms, and running experiments on quantum hardware. Qiskit also offers a rich set of libraries and tools for quantum machine learning, quantum chemistry, and quantum optimization. Another prominent quantum programming language is Cirq, developed by Google. Cirq is designed for programming quantum devices, with a focus on quantum circuits and gates. It offers a Python-based language for specifying quantum operations, circuits, and simulations. Cirq provides fine-grained control over quantum gates and allows developers to work directly with qubits, making it suitable for quantum hardware research and experimentation. Microsoft's Quantum Development Kit (Q#) is another quantum programming language that targets quantum algorithms and quantum simulations. Q# is designed for expressing quantum operations, quantum functions, and quantum data types. It integrates seamlessly with Visual Studio and offers quantum simulators for algorithm development and debugging. Q# also includes a resource estimation feature that helps developers analyze the resource requirements of quantum algorithms. Quipper, developed by Microsoft Research and the University of Oxford, is a quantum programming language designed for quantum circuit description and optimization. Quipper uses a functional programming style and offers a high-level language for specifying quantum circuits. It includes a compiler that optimizes quantum circuits for different quantum hardware architectures. A unique aspect of Quipper is its ability to perform automatic circuit optimization, enhancing the efficiency of quantum algorithms. Silq is a quantum programming language developed at ETH Zurich, designed to address the challenge of error-prone quantum computations. Silq introduces a concept called "error avoidance," where the language enforces correctness by design. It helps quantum developers write code that avoids common quantum programming pitfalls, such as uninitialized qubits or non-unitary operations. Silq's type system ensures that quantum programs adhere to quantum mechanics principles, reducing the potential for errors. Strawberry Fields is a quantum programming framework developed by Xanadu that focuses on photonic quantum computing. It uses a Python-based quantum programming language to specify quantum algorithms and quantum circuits using continuous-variable quantum systems. Strawberry Fields integrates with quantum hardware platforms, such as photonic quantum processors, and offers quantum machine learning libraries. The quantum programming landscape also includes domain-specific languages tailored for quantum algorithms in chemistry, optimization, and cryptography. For example, OpenFermion, an open-source platform, provides a quantum programming language for quantum chemistry simulations. OpenFermion allows researchers to express molecular Hamiltonians and perform quantum simulations to study molecular properties. Quantum programming languages for optimization, such as QAOA, provide abstractions for solving combinatorial optimization problems on quantum hardware. These languages enable the formulation of optimization problems and the execution of quantum algorithms to find optimal solutions. In the realm of quantum cryptography, languages like QKD (Quantum Key Distribution) languages are used to define protocols for secure quantum communication. Advanced quantum programming languages aim to address the challenges of quantum programming, such as qubit management, quantum error correction, and quantum resource optimization. Quantum programming languages are designed to hide the complexity of quantum hardware and provide high-level abstractions for quantum software developers. They offer tools for composing quantum algorithms, simulating quantum circuits, and interfacing with quantum hardware. Quantum software development often involves a combination of quantum and classical languages. Classical languages, such as Python or C++, are used for control and classical post-processing of quantum computations. Quantum programming languages, on the other hand, are responsible for specifying the quantum operations and logic. Interoperability between quantum and classical languages is a critical aspect of quantum software development. Advanced quantum programming languages continue to evolve and expand their capabilities, driven by both research institutions and industry leaders. These languages play a pivotal role in advancing quantum computing, enabling the development of quantum algorithms and applications across various domains. Quantum software development is a dynamic and interdisciplinary field that combines expertise in quantum physics, computer science, and mathematics. As quantum hardware continues to advance, quantum programming languages will play a crucial role in harnessing the potential of quantum computing to solve real-world problems and drive

scientific discovery.
Quantum software development tools are essential resources for harnessing the power of quantum computing and creating quantum applications. These tools provide the necessary infrastructure for quantum software developers to design, simulate, and run quantum algorithms on various quantum hardware platforms. Quantum software development tools encompass a wide range of software libraries, frameworks, and simulators that facilitate the creation and execution of quantum programs. One of the fundamental elements of quantum software development tools is the quantum programming language, which serves as the interface for expressing quantum algorithms. These languages, such as Qiskit, Cirq, and Q#, provide abstractions and syntax for specifying quantum circuits, gates, and quantum operations. They enable developers to design quantum algorithms by composing a sequence of quantum gates and operations that manipulate qubits to perform specific tasks. Quantum programming languages are typically designed to be user-friendly, allowing developers to focus on algorithm design rather than the intricacies of quantum hardware. Another crucial component of quantum software development tools is the quantum simulator, which enables the execution and testing of quantum programs on classical computers. Simulators mimic the behavior of quantum hardware and provide a controlled environment for debugging and verifying quantum algorithms. Quantum simulators are invaluable for quantum software developers as they allow for efficient algorithm development and testing before running on real quantum hardware. Major quantum programming frameworks, like Qiskit and Cirq, offer built-in simulators that facilitate quantum algorithm simulation and debugging. In addition to quantum simulators, quantum software development tools often include resource estimation features that help developers analyze the computational resources required for running

quantum algorithms on quantum hardware. Resource estimation is crucial for assessing the feasibility and efficiency of quantum algorithms. Quantum software development tools are designed to be platform-agnostic, allowing developers to work with various quantum hardware backends. These backends can include superconducting qubits, trapped ions, photonic qubits, and other quantum technologies. Quantum software developers can choose the backend that best suits their needs and preferences. Quantum software development tools provide a seamless interface for executing quantum programs on different quantum hardware platforms, abstracting away the platform-specific details. Quantum software development tools are continuously evolving to keep pace with advancements in quantum hardware. They are developed and maintained by organizations, including major tech companies, academic institutions, and quantum startups. These organizations are heavily invested in quantum software development, as they recognize the pivotal role that software plays in realizing the potential of quantum computing. Cloud-based quantum computing platforms, such as IBM Quantum Experience and Google Quantum Cloud, offer developers access to quantum hardware through quantum software development tools. These platforms provide quantum hardware resources via the cloud, allowing developers to run quantum programs remotely. Quantum software development tools enable users to access quantum hardware resources, submit quantum jobs, and retrieve results seamlessly. Quantum cloud platforms democratize access to quantum computing, making it available to a broader community of researchers, developers, and organizations. Quantum software development tools also play a significant role in quantum algorithm development and optimization. Quantum algorithms are designed to solve specific problems more efficiently than classical algorithms, but their development often involves a deep understanding of quantum mechanics and careful optimization. Quantum *software development tools provide tools and libraries for implementing quantum algorithms, as well as for optimizing and benchmarking them. Quantum algorithm development is an interdisciplinary field that combines quantum physics, computer science, and mathematics. Quantum software developers collaborate with quantum hardware engineers and researchers to ensure that quantum algorithms are tailored to the capabilities and constraints of quantum hardware. Quantum software development tools are also crucial for quantum machine learning, a rapidly growing area of research. Quantum machine learning algorithms leverage quantum properties to perform tasks such as data classification, optimization, and linear algebra more efficiently than classical counterparts. Quantum software developers use quantum software development tools to implement and test these quantum machine learning algorithms on quantum hardware. Quantum software development tools are part of a vibrant ecosystem of quantum technologies and research. They enable the exploration of quantum computing's potential in fields such as cryptography, optimization, material science, and drug discovery. Quantum software development tools are central to realizing the practical applications of quantum computing in various industries. Quantum software development is not limited to experienced quantum physicists but is accessible to a broader audience, including software engineers, data scientists, and researchers from various domains. Quantum software development tools offer a bridge between classical and quantum computing, empowering developers to leverage the capabilities of quantum hardware without requiring expertise in quantum mechanics. As quantum hardware continues to advance, quantum software development tools will play a pivotal role in unlocking the potential of quantum computing to address real-world challenges and drive scientific innovation.*

Chapter 9: Quantum Cryptography and Quantum Communication

Advanced quantum cryptographic protocols are at the forefront of securing communication and information exchange in the quantum age. These protocols leverage the unique properties of quantum mechanics to provide unprecedented levels of security and privacy. Quantum cryptography represents a paradigm shift in secure communication, offering solutions that are fundamentally different from classical cryptographic methods. One of the foundational concepts in quantum cryptography is quantum key distribution (QKD), which enables two parties to establish a secret cryptographic key with provable security guarantees. QKD protocols, such as the BBM92 (Bennett-Brassard 1992) protocol, use quantum properties like the no-cloning theorem and the uncertainty principle to ensure that any eavesdropping attempts are detectable. Advanced QKD protocols have been developed to enhance the practicality and security of quantum key distribution. For instance, the E91 protocol introduced the idea of entanglement-based QKD, where entangled particles are used to create the secret key. Entanglement-based QKD offers advantages in terms of key generation rates and resistance against certain attacks. Additionally, the development of QKD systems that can operate over long distances has expanded the scope of secure communication. Technologies like quantum repeaters and satellite-based QKD enable secure key distribution over global distances. Another quantum cryptographic protocol gaining attention is quantum secure direct communication (QSDC). Unlike QKD, which focuses on key distribution, QSDC allows two parties to exchange messages directly in a secure and confidential manner. Quantum teleportation plays a crucial role in QSDC protocols, ensuring that the transmitted information is inaccessible to eavesdroppers. Furthermore, quantum secure direct communication can be implemented using various quantum systems, such as superconducting qubits, trapped ions, and photonic qubits. Quantum digital signatures represent another advancement in quantum cryptography. Digital signatures are essential for verifying the authenticity and integrity of digital messages. Quantum digital signatures use quantum properties to provide unforgeable signatures that cannot be duplicated or tampered with. These signatures rely on the properties of quantum states, such as the no-cloning theorem, to ensure security. Quantum-resistant cryptography is an emerging field that addresses the threat posed by quantum computers to classical cryptographic algorithms. Quantum computers have the potential to break widely used encryption schemes, such as RSA and ECC, by exploiting their computational power to solve certain mathematical problems efficiently. To counter this threat, post-quantum cryptography research is focused on developing cryptographic algorithms that are secure against attacks by quantum computers. Lattice-based cryptography, code-based cryptography, and hash-based cryptography are among the candidates for post-quantum security. Quantum-resistant cryptographic protocols aim to protect sensitive information in a quantum-safe manner, ensuring that data remains secure even in the presence of powerful quantum adversaries. Quantum-resistant algorithms are designed to be secure against both classical and quantum attacks, making them a crucial component of advanced quantum cryptography. Quantum-resistant cryptography is becoming increasingly important as quantum computing technology advances. Quantum-resistant cryptographic protocols are essential for safeguarding data and communications in a world where quantum computers may pose a significant threat to classical cryptographic systems. Quantum-resistant encryption schemes, digital signatures, and key exchange protocols provide security guarantees that extend beyond the capabilities of classical cryptographic techniques. Quantum-resistant cryptography is a multidisciplinary field that brings together experts in mathematics, computer science, and quantum physics to design algorithms that resist attacks from both classical and quantum adversaries. Quantum-resistant cryptographic protocols are being standardized and adopted by organizations and industries to prepare for the *post-quantum era. The development and deployment of quantum-resistant cryptography are essential steps in ensuring the long-term security of digital communication and data protection. Quantum-resistant encryption algorithms, such as NTRUEncrypt and Ring-LWE-based schemes, provide a robust defense against quantum attacks. These algorithms are designed to withstand attacks from quantum computers while offering the same level of security as classical encryption. Quantum-resistant digital signatures, such as the Dilithium and Falcon schemes, enable secure authentication and message integrity in a quantum-safe manner. These signatures are resistant to attacks by quantum adversaries and provide long-term security for digital signatures. Quantum-resistant key exchange protocols, such as Kyber and NewHope, allow secure establishment of shared cryptographic keys even in the presence of quantum computers. These protocols ensure that confidential information remains protected in a post-quantum world. Quantum-resistant cryptography is a rapidly evolving field, with ongoing research and development to address emerging challenges and threats. Researchers and organizations are working to standardize quantum-resistant cryptographic algorithms to ensure widespread adoption and integration into existing security protocols. Quantum-resistant cryptography is a critical component of advanced quantum cryptographic protocols, safeguarding sensitive information and ensuring the long-term security of digital communication. As quantum technologies continue to advance, quantum-resistant cryptography will play a vital role in protecting data and maintaining the confidentiality and integrity of digital transactions. Secure quantum communication networks represent a significant advancement in the field of secure communications, offering unprecedented levels of security and privacy. These networks harness the principles of quantum mechanics to protect information during transmission, ensuring that it remains confidential and immune to eavesdropping attempts. Quantum communication relies on the properties of quantum states, such as superposition and entanglement, to achieve security guarantees that classical communication methods cannot provide. One of the key applications of secure quantum communication is quantum key distribution (QKD), which allows two parties to establish a secret cryptographic key with a level of security that is theoretically unbreakable. QKD protocols, such as the BBM92 (Bennett-Brassard 1992) protocol, use quantum properties to ensure that any eavesdropping attempts are detectable, making it impossible for an adversary to obtain the secret key without detection. Secure quantum communication networks leverage QKD protocols to create secure communication channels, making them impervious to interception or decryption by malicious actors. The development of secure quantum communication networks involves the deployment of quantum key distribution systems over various physical mediums. Quantum key distribution systems can operate over optical fibers, free-space optical links, and even satellite-based connections. Optical fiber-based QKD systems are widely deployed in metropolitan and long-distance communication networks. These systems use quantum properties of photons, such as polarization or phase encoding, to transmit cryptographic keys securely over long distances. Free-space optical links are employed for secure communication between ground stations and satellites. Satellite-based QKD allows for global key distribution and secure communication between geographically separated locations. Secure quantum communication networks provide a secure foundation for various applications, including secure video conferencing, financial transactions, and government communications. They protect sensitive information from potential cyberattacks and data breaches, ensuring the confidentiality and integrity of communication. Secure quantum communication networks also have applications in secure voting systems, enabling secure and verifiable electronic voting while preserving voter anonymity. Quantum teleportation plays a crucial role in secure quantum communication networks. Quantum teleportation allows the transmission of an unknown quantum state from one location to another without physically moving the quantum particles themselves. This property enables the secure transfer of quantum cryptographic keys, ensuring that they remain confidential and intact during transmission. Quantum teleportation is a fundamental building block of quantum repeaters, which are essential for extending the range of secure quantum communication networks. Quantum repeaters are devices that can regenerate and amplify quantum signals, allowing secure key distribution over long distances. They are crucial for global-scale secure quantum communication networks and are an active area of research and development. Secure quantum communication networks also incorporate quantum-resistant cryptography to protect against potential threats from quantum computers. Quantum computers have the potential to break classical cryptographic algorithms, such as RSA and ECC, by solving certain mathematical problems efficiently. To counter this threat, quantum-resistant cryptography is integrated into secure quantum communication networks. Post-quantum cryptographic algorithms, such as lattice-based and code-based cryptography, provide security against both classical and quantum attacks. Quantum-resistant cryptography ensures that the cryptographic keys and data exchanged in secure quantum communication networks remain secure even in a future where quantum computers are prevalent. One of the challenges in the deployment of secure quantum communication networks is the integration of quantum technology with existing classical infrastructure. Quantum key distribution systems must be seamlessly integrated into classical communication networks, ensuring compatibility and reliability. Quantum network hardware, such as quantum repeaters and quantum routers, must be designed to work alongside classical network components. Efforts are underway to standardize and develop protocols for the seamless integration of quantum technology into existing communication networks. Secure quantum communication networks also require advanced error correction techniques to mitigate the effects of quantum noise and errors in transmission. Quantum error correction codes, such as the surface code, are used to protect quantum information from errors during key distribution. These codes rely on syndrome measurements to identify errors and apply corrective operations to recover the quantum state. Efforts are ongoing to improve the efficiency and resource overhead of quantum error correction in secure quantum communication networks. Secure quantum communication networks are a multidisciplinary field that combines expertise in quantum physics, computer science, networking, and cryptography. Researchers and engineers work collaboratively to develop and deploy secure quantum communication technologies. Major advancements in secure quantum communication networks are driven by collaborations between governments, research institutions, and private industry. National and international initiatives aim to develop and deploy secure quantum communication networks to protect critical infrastructure, government communications, and sensitive information. The development and deployment of secure quantum communication networks represent a significant step towards a future where secure communication is both achievable and scalable on a global scale. As quantum technology continues to advance, secure quantum communication networks will play a pivotal role in ensuring the security and privacy of digital communication in an increasingly interconnected world.*

Chapter 10: Quantum Computing in Practice: Real-World Applications

Real-world quantum computing use cases showcase the transformative potential of quantum technologies across various domains and industries. Quantum computing is not just a theoretical concept; it is a rapidly evolving field that is beginning to have a tangible impact on practical applications. One of the most prominent and promising use cases for quantum computing is in the field of optimization. Quantum algorithms, such as the Quantum Approximate Optimization Algorithm (QAOA), have demonstrated the ability to solve complex optimization problems more efficiently than classical algorithms. This has practical applications in supply chain management, portfolio optimization, and route planning, where finding optimal solutions quickly can lead to significant cost savings and improvements in efficiency. In the realm of material science, quantum computing holds the potential to revolutionize the discovery and development of new materials with desirable properties. Simulating the behavior of molecules and materials at the quantum level is a computationally intensive task, and quantum computers can perform these simulations more accurately and quickly than classical computers. This has applications in designing novel catalysts for chemical processes, developing advanced materials for electronics, and optimizing drug molecules for pharmaceuticals. Quantum machine learning is another emerging use case that leverages the power of quantum computing to enhance machine learning algorithms. Quantum machine learning algorithms can process and analyze large datasets more efficiently, enabling improvements in data classification, clustering, and predictive modeling. These advancements are particularly valuable in fields like finance, healthcare, and recommendation systems, where data-driven decisions are critical. Quantum cryptography represents a real-world use case that addresses the growing threat posed by quantum computers to classical encryption schemes. Quantum key distribution (QKD) protocols, such as the BBM92 (Bennett-Brassard 1992) protocol, enable secure communication by allowing two parties to establish a secret cryptographic key that is theoretically unbreakable, even by quantum computers. This technology is being deployed in secure communication networks and financial institutions to protect sensitive data from potential quantum attacks. Quantum chemistry is an area where quantum computing can make a significant impact. Modeling complex quantum systems, such as molecular structures and chemical reactions, is a computationally intensive task that often exceeds the capabilities of classical supercomputers. Quantum computers can provide accurate simulations of quantum chemistry problems, which have applications in drug discovery, materials science, and environmental research. In the field of artificial intelligence, quantum computing is poised to accelerate the training of machine learning models. Quantum-enhanced algorithms can potentially reduce the time required to train deep neural networks and improve the optimization process. This can lead to advancements in natural language processing, image recognition, and autonomous systems. Quantum cryptography is being used to secure data transmission in various industries, including finance, healthcare, and government communications. Quantum-resistant cryptographic protocols are implemented to protect sensitive information from potential threats posed by quantum computers. These protocols ensure that data remains secure in a post-quantum world, where classical *encryption schemes may be vulnerable. Quantum-enhanced financial modeling is revolutionizing risk assessment and investment strategies in the finance industry. Quantum algorithms can perform complex financial simulations and portfolio optimization more efficiently, enabling better-informed investment decisions and risk management. Quantum-enhanced optimization is making supply chain management more efficient by solving complex logistical problems in real-time. This has applications in reducing transportation costs, minimizing inventory, and optimizing the allocation of resources. Quantum-enhanced drug discovery is accelerating the development of new pharmaceuticals by simulating molecular interactions and identifying potential drug candidates more quickly and accurately. This has the potential to revolutionize the healthcare industry by bringing life-saving drugs to market faster. Quantum-enhanced materials science is driving the development of advanced materials with unique properties, such as superconductors and high-temperature superconductors, which have applications in energy transmission and storage. Quantum-enhanced logistics and route planning are optimizing delivery routes for shipping companies, reducing fuel consumption and emissions. Quantum-enhanced recommendation systems are improving the accuracy of product recommendations for e-commerce platforms, enhancing the customer experience. Quantum-enhanced data analysis is speeding up data processing and analysis, enabling businesses to extract valuable insights from large datasets more efficiently. Quantum-enhanced artificial intelligence is enhancing machine learning algorithms, enabling faster and more accurate training of neural networks. Quantum-enhanced natural language processing is improving language translation, sentiment analysis, and chatbot interactions. Quantum-enhanced autonomous systems are enhancing the capabilities of autonomous vehicles, drones, and robotics by improving perception and decision-making. Quantum-enhanced cybersecurity is strengthening the defenses of critical infrastructure, financial institutions, and government agencies against cyberattacks. Quantum-enhanced weather forecasting is improving the accuracy of weather predictions, helping mitigate the impact of natural disasters. Quantum-enhanced environmental modeling is enabling more accurate simulations of climate change and its effects on ecosystems. Quantum-enhanced energy optimization is improving the efficiency of energy production and distribution, reducing carbon emissions. Quantum-enhanced space exploration is advancing our understanding of the cosmos and enabling more efficient space missions. Real-world quantum computing use cases span a wide range of industries and applications, demonstrating the transformative potential of quantum technologies. As quantum hardware continues to advance, these use cases will become even more impactful, reshaping industries, solving complex problems, and pushing the boundaries of what is possible in the world of computation and science.*

Industry-specific applications of quantum computing are revolutionizing how various sectors tackle complex problems and drive innovation. Quantum computing is not a one-size-fits-all solution but a versatile tool with the potential to transform specific industries. In healthcare, quantum computing holds great promise for drug discovery and molecular modeling. By simulating the behavior of molecules and proteins at the quantum level, quantum computers can significantly accelerate the process of identifying potential drug candidates and understanding their interactions with biological systems. This has the potential to revolutionize the pharmaceutical industry, bringing life-saving drugs to market faster and more efficiently. In finance, quantum computing is enhancing risk management and portfolio optimization. Quantum algorithms can analyze vast amounts of financial data and perform complex simulations to identify investment opportunities and mitigate risks. This is particularly valuable in a highly competitive and data-driven industry where informed decisions can make a significant difference. In the energy sector, quantum computing is being applied to optimize energy production and distribution. Quantum algorithms can tackle complex optimization problems related to grid management, renewable energy integration, and resource allocation, leading to more efficient and sustainable energy systems. In aerospace and defense, quantum computing is revolutionizing complex simulations and modeling. From aerodynamics and materials science to cryptography and threat analysis, quantum computers are enabling more accurate and faster simulations that enhance the capabilities of military and aerospace applications. In logistics and transportation, quantum computing is transforming route optimization and supply chain management. By solving intricate logistical problems in real-time, quantum algorithms can reduce transportation costs, minimize inventory, and improve resource allocation, benefiting shipping companies and the global supply chain. In the automotive industry, quantum computing is advancing the design and optimization of vehicle systems. From aerodynamics and safety simulations to battery optimization and traffic management, quantum computers are driving innovation in automotive engineering, making vehicles more efficient and sustainable. In telecommunications, quantum computing is being used to improve network optimization and data security. Quantum algorithms can efficiently solve network routing problems and enhance the encryption of sensitive data, safeguarding communications in an increasingly connected world. In materials science, quantum computing is accelerating the discovery of advanced materials with unique properties. Simulating the behavior of molecules and materials at the quantum level enables researchers to design novel materials for applications in electronics, energy storage, and nanotechnology. In environmental science, quantum computing is aiding in climate modeling and ecosystem analysis. Advanced simulations powered by quantum computers provide more accurate insights into climate change, helping researchers understand its impact and develop strategies for mitigation. In agriculture, quantum computing is optimizing crop management and resource allocation. By analyzing agricultural data and environmental factors, quantum algorithms can assist farmers in making informed decisions about planting, irrigation, and pest control, ultimately improving crop yield and sustainability. In the pharmaceutical industry, quantum computing is revolutionizing drug discovery and molecular modeling. By simulating the interactions of molecules and proteins at the quantum level, quantum computers can significantly accelerate the process of identifying potential drug candidates and understanding their mechanisms of action. This has the potential to bring new medicines to market more quickly and efficiently. In the retail sector, quantum computing is enhancing supply chain management and demand forecasting. Quantum algorithms can optimize inventory management, transportation logistics, and pricing strategies, improving efficiency and customer satisfaction. In the entertainment industry, quantum computing is advancing content creation and distribution. From optimizing streaming services and rendering visual effects to enhancing recommendation systems and personalized content delivery, quantum technology is transforming the entertainment *landscape. In the legal profession, quantum computing is revolutionizing document analysis and legal research. Quantum algorithms can sift through vast legal databases, extract relevant information, and assist in legal research and case analysis, streamlining legal processes. In the insurance industry, quantum computing is improving risk assessment and underwriting. Quantum algorithms can process extensive datasets and perform complex simulations to evaluate risks and determine insurance pricing more accurately, benefiting both insurers and policyholders. In the education sector, quantum computing is fostering innovation in research and educational technology. Quantum simulations and modeling enhance scientific research across various disciplines, while quantum-enhanced educational tools provide new ways to teach complex concepts and problem-solving skills. In the entertainment industry, quantum computing is being used to create more immersive virtual reality experiences and optimize content delivery to consumers. Quantum algorithms can enhance the rendering of 3D graphics and simulations, creating more realistic and engaging virtual worlds. In the hospitality industry, quantum computing is revolutionizing hotel and restaurant management. Quantum algorithms can optimize booking systems, resource allocation, and customer experience, improving operational efficiency and guest satisfaction. In the gaming industry, quantum computing is driving advancements in game design and artificial intelligence. Quantum-enhanced algorithms can create more complex and challenging gameplay experiences, as well as improve character behavior and decision-making in games. In the automotive industry, quantum computing is advancing the development of autonomous vehicles and improving traffic management. Quantum algorithms can optimize route planning, traffic flow, and vehicle control systems,* contributing to safer and more efficient transportation solutions. In the aerospace industry, quantum computing is revolutionizing aircraft design and aerodynamics. Quantum simulations enable more accurate modeling of complex fluid dynamics and materials, leading to the development of more fuel-efficient and environmentally friendly aircraft. In the pharmaceutical industry, quantum computing is accelerating drug discovery and personalized medicine. Quantum algorithms can analyze genetic data and simulate drug interactions, leading to more effective treatments and tailored medical solutions for patients. In the legal profession, quantum computing is transforming contract analysis and legal research. Quantum-enhanced algorithms can quickly analyze and extract critical information from legal documents, streamlining contract review and legal investigations. In the healthcare industry, quantum computing is advancing medical imaging and drug discovery. Quantum algorithms can process and analyze large medical datasets, leading to more accurate diagnoses and the discovery of novel treatments. These examples illustrate the diverse range of industry-specific applications of quantum computing, each addressing unique challenges and opportunities within their respective sectors. As quantum technology continues to advance, these applications will likely expand and evolve, driving innovation and reshaping industries in ways that were once thought impossible.

BOOK 3
ADVANCED QUANTUM COMPUTING
EXPLORING THE FRONTIERS OF COMPUTER SCIENCE,
PHYSICS, AND MATHEMATICS
ROB BOTWRIGHT

Chapter 1: Quantum Mechanics Foundations for Advanced Computing

Advanced concepts in quantum mechanics delve into the profound and intricate nature of the quantum world, challenging our understanding of reality. Quantum mechanics, developed in the early 20th century, has become one of the most successful and fundamental theories in physics. At its core, quantum mechanics describes the behavior of matter and energy at the smallest scales, where classical physics fails to provide accurate predictions. One of the foundational principles of quantum mechanics is the wave-particle duality, which asserts that particles like electrons and photons exhibit both wave-like and particle-like properties. This dual nature is encapsulated in the famous Schrödinger equation, which describes the evolution of quantum wave functions and governs the behavior of quantum systems. The wave function represents the probability amplitude of finding a particle in a particular state or position. One of the striking consequences of quantum mechanics is the uncertainty principle, formulated by Werner Heisenberg. The uncertainty principle states that there is a fundamental limit to the precision with which certain pairs of complementary properties, such as position and momentum, can be simultaneously known. This inherent uncertainty challenges our classical intuition and sets a fundamental bound on our knowledge of the quantum world. Quantum entanglement is a phenomenon that lies at the heart of many advanced quantum concepts. When two or more particles become entangled, their quantum states become correlated in such a way that the properties of one particle are instantaneously linked to the properties of another, even when separated by vast distances. This non-local connection, famously referred to as "spooky action at a distance" by Albert

Einstein, has been experimentally verified and is a central aspect of quantum mechanics. Quantum superposition is another key concept, illustrating that quantum particles can exist in multiple states simultaneously. For example, a quantum bit or qubit can represent both 0 and 1 at the same time, offering the potential for quantum computers to perform parallel computations. Quantum tunneling is a fascinating quantum phenomenon where particles can pass through energy barriers that would be insurmountable according to classical physics. This effect is responsible for phenomena like nuclear fusion in stars and the operation of tunnel diodes in electronics. Quantum decoherence is a challenge in quantum systems, where interactions with the environment can disrupt quantum superposition and entanglement, leading to a transition from the quantum to classical behavior. Understanding and mitigating decoherence is crucial for the development of quantum technologies. Quantum measurement is a fundamental process in quantum mechanics that collapses a quantum system's wave function to a specific state or outcome. The outcome of a quantum measurement is probabilistic, and the process itself remains a subject of philosophical debate. The Copenhagen interpretation of quantum mechanics, formulated by Niels Bohr and Werner Heisenberg, provides one perspective on the role of measurement in quantum theory. It suggests that the act of measurement causes the wave function to collapse into a definite state. The Many-Worlds Interpretation, proposed by Hugh Everett III, offers a different perspective on quantum measurement. In this interpretation, every possible outcome of a quantum measurement actually occurs in separate, non-communicating branches of the universe, creating a multiverse of parallel realities. Quantum entanglement can be harnessed for applications such as quantum teleportation and quantum cryptography. Quantum teleportation allows for the transmission of an unknown quantum state from one location to another by entangling the two locations. Quantum cryptography, including quantum key distribution, exploits the non-local properties of entangled particles to create secure communication channels that are theoretically immune to eavesdropping. Bell's theorem, formulated by John Bell, offers a way to test the predictions of quantum mechanics against those of classical physics and determine the presence of quantum entanglement. Experiments based on Bell's theorem have consistently shown results that support the predictions of quantum mechanics and confirm the non-classical nature of entanglement. Quantum states can be manipulated using quantum gates and operations, similar to classical logic gates in computing. Quantum gates, such as the Hadamard gate and CNOT gate, are the building blocks of quantum circuits. These gates enable the creation of quantum algorithms, which can solve certain problems exponentially faster than classical algorithms. Quantum algorithms like Shor's algorithm and Grover's algorithm have the potential to disrupt fields like cryptography and optimization. Shor's algorithm, for example, can factor large numbers exponentially faster than the best-known classical algorithms, threatening widely used encryption methods. Quantum computing hardware has advanced significantly in recent years. Different physical implementations of quantum bits include superconducting qubits, trapped ions, photonic qubits, and topological qubits. Each of these qubit types has its advantages and challenges, contributing to the diversity of quantum hardware research. Quantum error correction is a crucial area of study for building fault-tolerant quantum computers. Quantum error correction codes, such as the surface code, provide a means to detect and correct errors that can occur during quantum computations. Efforts in quantum error correction are essential for realizing practical and scalable quantum computing. Quantum simulations leverage quantum computers to model and understand complex quantum systems, such as molecules and materials. These simulations can offer insights into quantum chemistry, condensed matter physics, and other fields that involve quantum interactions. Quantum annealing, as demonstrated by D-Wave Systems, is an approach to optimization that leverages quantum effects to find low-energy solutions to complex problems. Quantum annealers have shown promise in tackling optimization problems in fields like finance and logistics. Quantum-enhanced sensors, such as quantum gravimeters and quantum magnetometers, take advantage of quantum properties to achieve unprecedented levels of sensitivity and precision. These sensors have applications in geophysics, mineral exploration, and navigation. Quantum communication networks are being developed to provide secure communication channels that are immune to eavesdropping. Quantum key distribution (QKD) systems enable secure key exchange between two parties, ensuring the confidentiality of transmitted data. Quantum-enhanced imaging techniques, such as quantum-enhanced lidar and quantum-enhanced microscopy, offer improved imaging capabilities for applications in remote sensing and biological research. Quantum-enhanced artificial intelligence is a rapidly growing field that explores how quantum computing can enhance machine learning algorithms and accelerate artificial intelligence applications. Quantum machine learning algorithms aim to improve data analysis and decision-making across various domains. Quantum-enhanced cryptography is a critical area of research that addresses the security challenges posed by quantum computers. Post-quantum cryptographic protocols and quantum-resistant algorithms are being developed to safeguard sensitive information from potential quantum attacks. These advanced concepts in quantum mechanics highlight the rich and diverse landscape of quantum physics, from fundamental principles like wave-particle duality and the uncertainty principle to practical applications like quantum computing, quantum communication, and quantum-enhanced technologies. As our understanding of quantum mechanics deepens and quantum *technologies continue to advance, the potential for new discoveries and innovations in the quantum realm remains limitless.*

Quantum mechanics, a foundational theory in physics, has ushered in a new era of computing with applications that challenge the boundaries of classical computation. Quantum computing harnesses the unique principles of quantum mechanics to perform certain calculations at speeds unattainable by classical computers. At the heart of quantum computing are quantum bits or qubits, which differ fundamentally from classical bits. While classical bits can only be in a state of 0 or 1, qubits can exist in a superposition of both states simultaneously. This property enables quantum computers to explore multiple solutions to a problem in parallel, exponentially increasing their computational power. Quantum computing has garnered significant attention for its potential to solve complex problems in fields such as cryptography, optimization, and materials science. One of the most well-known quantum algorithms is Shor's algorithm, developed by Peter Shor in 1994. Shor's algorithm is designed to factor large numbers exponentially faster than the best-known classical algorithms. This capability poses a significant threat to widely used encryption methods, as the security of many cryptographic systems relies on the difficulty of factoring large numbers. Quantum computers could potentially break these encryption schemes, necessitating the development of quantum-resistant cryptography. Another groundbreaking quantum algorithm is Grover's algorithm, proposed by Lov Grover in 1996. Grover's algorithm significantly accelerates the search for a specific item within an unsorted database. Classically, this task requires checking each item one by one, which scales linearly with the size of the database. In contrast, Grover's algorithm provides a quadratic speedup, making it valuable for tasks like unstructured search and database querying. Quantum computing's potential for optimization is also of great interest in various industries. The Quantum Approximate Optimization Algorithm (QAOA) is a quantum algorithm designed to solve complex optimization problems. These problems can be found in logistics, finance, drug discovery, and supply chain management, among others. QAOA leverages quantum mechanics to find approximate solutions to optimization problems more efficiently than classical algorithms. Quantum computing's impact extends beyond solving specific problems; it also has the potential to revolutionize machine learning. Quantum machine learning algorithms leverage the power of quantum computers to improve data analysis, classification, and predictive modeling. These algorithms can process and analyze large datasets more efficiently, making them valuable in fields where data-driven decisions are crucial, such as finance, healthcare, and recommendation systems. Quantum machine learning holds the promise of creating more accurate models and accelerating the development of artificial intelligence. Quantum cryptography is another application of quantum mechanics that addresses the growing threat posed by quantum computers to classical encryption schemes. Quantum key distribution (QKD) protocols, such as the BBM92 protocol developed by Charles Bennett and Gilles Brassard in 1992, enable secure communication by allowing two parties to establish a secret cryptographic key that is theoretically unbreakable, even by quantum computers. This technology is being deployed in secure communication networks and financial institutions to protect sensitive data from potential quantum attacks. Quantum cryptography offers a fundamentally secure way to transmit information, as any attempt to eavesdrop on the communication would disrupt the quantum state and be detectable. Quantum computing's applications in materials science are also transformative. Quantum computers can simulate and analyze the behavior of molecules and materials at the quantum level with unprecedented accuracy. This capability has applications in designing novel catalysts for chemical processes, developing advanced materials for electronics, and optimizing drug molecules for pharmaceuticals. Researchers can use quantum simulations to understand molecular interactions and predict the properties of materials, opening up new possibilities for scientific discovery and innovation. Quantum computing's potential in the field of artificial intelligence is not limited to machine learning algorithms. Quantum-enhanced optimization can also improve the training of deep neural networks, speeding up the optimization process. This advancement can lead to more efficient natural language processing, image recognition, and autonomous systems. Quantum computing's implications for cryptography extend to post-quantum cryptography. As quantum computers advance, they pose a significant threat to classical encryption methods, which rely on the difficulty of solving certain mathematical problems. Post-quantum cryptography is a field of research focused on developing cryptographic algorithms that are secure against quantum attacks. These algorithms aim to protect sensitive information in a quantum-safe manner, ensuring that data remains secure even in a future where quantum computers are prevalent. Quantum-resistant encryption schemes, digital signatures, and key exchange protocols provide security guarantees that extend beyond the capabilities of classical cryptographic techniques. Quantum computing's impact on optimization is also relevant in finance, where efficient portfolio optimization and risk assessment are critical. Quantum algorithms can process vast amounts of financial data and perform complex simulations, enabling more informed investment decisions and risk management. In the energy sector, quantum computing is being applied to optimize energy production, distribution, and resource allocation. Quantum algorithms can solve complex optimization problems related to grid management, renewable energy integration, and resource allocation, leading to more efficient and sustainable energy systems. Quantum computing's role in aerospace and defense is multifaceted. It can improve the design of aircraft, optimize military logistics, and enhance threat analysis by providing more accurate simulations and modeling. Quantum-enhanced sensors, such as quantum gravimeters and quantum magnetometers, can enhance military capabilities in navigation, surveillance, and reconnaissance. Quantum computing also offers new possibilities in logistics and transportation. By solving complex logistical problems in real-time, quantum algorithms can reduce transportation costs, minimize inventory, and optimize the allocation of resources. Quantum-enhanced sensors can provide more accurate information for autonomous vehicles and drones, enhancing their safety and efficiency. In agriculture, quantum computing is optimizing crop management and resource allocation. By analyzing agricultural data and environmental factors, quantum algorithms can assist farmers in making informed decisions about planting, irrigation, and pest control, ultimately improving crop yield and sustainability. Quantum computing's potential applications are not limited to the aforementioned areas. It has the potential to revolutionize various other fields, including telecommunications, retail, entertainment, and environmental science. Quantum computing holds the promise of tackling complex problems, accelerating scientific discovery, and reshaping industries across the globe. As quantum technology continues to advance, its real-world applications will become increasingly prevalent, offering innovative solutions to some of the world's most challenging problems.

Chapter 2: Quantum Information Theory and Quantum Entanglement

Quantum information theory, an intriguing and rapidly evolving field, delves into advanced topics that explore the profound connections between quantum mechanics and information processing. Building on the foundational principles of quantum mechanics, quantum information theory investigates how quantum systems can be harnessed to store, transmit, and process information in ways that challenge classical paradigms. Entanglement, a key concept in quantum information theory, lies at the heart of many advanced topics. Entanglement occurs when two or more particles become correlated in such a way that their quantum states are intertwined, even when they are separated by vast distances. This non-local connection defies classical intuition and is a fundamental resource in quantum information processing. Quantum teleportation, a remarkable phenomenon enabled by entanglement, allows for the transfer of an unknown quantum state from one location to another. This process relies on entangled particles to transmit quantum information accurately and securely, paving the way for quantum communication and quantum networking. Quantum cryptography, a field closely related to quantum information theory, exploits the principles of quantum mechanics to create secure communication channels that are theoretically impervious to eavesdropping. Quantum key distribution (QKD) protocols, such as the BBM92 protocol developed by Charles Bennett and Gilles Brassard, establish secret cryptographic keys between two parties in a way that any interception of the key can be detected. This technology is being implemented in secure communication networks, financial institutions, and government agencies to safeguard sensitive information from potential quantum attacks. Quantum error correction, an essential component of quantum information theory, addresses the susceptibility of quantum systems to errors and decoherence. Quantum error correction codes, such as the surface code, provide a means to detect and correct errors that can occur during quantum computations. Efforts in quantum error correction are vital for building practical and scalable quantum computers that can perform complex calculations reliably. Quantum algorithms, a cornerstone of quantum information theory, offer the potential for exponential speedup in solving certain problems compared to classical algorithms. Shor's algorithm, for instance, can factor large numbers exponentially faster than the best-known classical algorithms, posing a significant threat to classical encryption methods. Grover's algorithm, on the other hand, accelerates database searching and unstructured search problems by providing a quadratic speedup. Quantum algorithms have far-reaching implications across various domains, from cryptography to optimization and machine learning. Quantum computing, a major application of quantum information theory, leverages quantum bits or qubits to perform calculations beyond the capabilities of classical computers. Quantum computers exploit quantum parallelism and entanglement to solve complex problems efficiently. While quantum hardware is still in its nascent stages, it holds the promise of revolutionizing fields such as materials science, drug discovery, and cryptography. Quantum communication networks, another practical application of quantum information theory, are emerging as a secure means of transmitting information. These networks rely on the principles of quantum key distribution (QKD) to establish secure communication channels that are resistant to quantum attacks. The development of quantum repeaters and quantum satellites is extending the reach of secure quantum communication over longer distances. Quantum-enhanced sensing technologies, such as quantum gravimeters and quantum magnetometers, leverage the principles of quantum mechanics to achieve unprecedented levels of sensitivity and precision. These sensors have applications in geophysics, mineral exploration, and navigation, offering valuable insights and improving the accuracy of measurements. Quantum-enhanced imaging techniques, such as quantum-enhanced lidar and quantum-enhanced microscopy, provide advanced imaging capabilities for applications in remote sensing and biological research. These techniques offer greater accuracy, resolution, and depth perception, leading to breakthroughs in scientific discovery and industrial applications. Quantum-enhanced metrology, a burgeoning field in quantum information theory, enhances the precision of measurements, including timekeeping and spectroscopy. Quantum clocks, which utilize the quantum properties of entangled particles, have the potential to redefine timekeeping standards with unprecedented accuracy. Quantum-enhanced cryptography is a critical area of research focused on developing cryptographic protocols that are secure against quantum attacks. These post-quantum cryptographic algorithms aim to protect sensitive information from potential threats posed by quantum computers. Quantum-resistant encryption schemes, digital signatures, and key exchange protocols are being designed to ensure the security of data in a post-quantum world. Quantum-enhanced algorithms in machine learning, a rapidly growing field, harness the power of quantum computers to improve data analysis and predictive modeling. Quantum machine learning algorithms aim to accelerate tasks such as data classification, clustering, and optimization, with applications spanning finance, healthcare, and recommendation systems. Quantum-enhanced artificial intelligence is a frontier in quantum information theory that seeks to enhance the training and performance of machine learning models. Quantum algorithms have the potential to reduce the time required for training deep neural networks and improve optimization processes, leading to advancements in natural language processing and computer vision. Quantum-

enhanced materials science is at the forefront of quantum information theory, revolutionizing the discovery and development of novel materials with exceptional properties. Quantum computers can simulate and analyze the behavior of molecules and materials at the quantum level, providing insights into materials for electronics, energy storage, and pharmaceuticals. Quantum-enhanced optimization, an application with wide-ranging impact, addresses complex optimization problems in fields such as logistics, finance, and supply chain management. Quantum algorithms like the Quantum Approximate Optimization Algorithm (QAOA) offer the potential to find optimal solutions more efficiently than classical algorithms. Quantum simulations, another application of quantum information theory, enable researchers to model and understand complex quantum systems, from molecules to materials. These simulations have applications in quantum chemistry, condensed matter physics, and quantum field theory, driving advancements in scientific understanding and technological innovation. Quantum-enhanced finance models are transforming risk assessment, portfolio optimization, and investment strategies in the financial industry. Quantum algorithms can perform complex financial simulations and provide more accurate insights into market trends and investment decisions. Quantum-enhanced logistics and supply chain management are optimizing resource allocation, transportation routes, and inventory management. This has implications for reducing costs, minimizing environmental impact, and improving the efficiency of global supply chains. Quantum-enhanced drug discovery is accelerating the development of new pharmaceuticals by simulating molecular interactions and identifying potential drug candidates more quickly and accurately. This has the potential to revolutionize healthcare by bringing life-saving drugs to market faster. Quantum-enhanced materials science is driving the development of advanced materials with unique properties, such as superconductors and high-temperature superconductors. These materials have applications in energy transmission and storage, medical devices, and electronics. Quantum-enhanced logistics and route planning are optimizing delivery routes for shipping companies, reducing fuel consumption and emissions. Quantum-enhanced recommendation systems are improving the accuracy of product recommendations for e-commerce platforms, enhancing the customer experience. Quantum-enhanced data analysis is speeding up data processing and analysis, enabling businesses to extract valuable insights from large datasets more efficiently. Quantum-enhanced artificial intelligence is enhancing machine learning algorithms, enabling faster and more accurate training of neural networks. Quantum-enhanced natural language processing is improving language translation, sentiment analysis, and chatbot interactions. Quantum-enhanced autonomous systems are enhancing the capabilities of autonomous vehicles, drones, and robotics by improving perception and decision-making. Quantum-enhanced cybersecurity is strengthening the defenses of critical infrastructure, financial institutions, and government agencies against cyberattacks. Quantum-enhanced weather forecasting is improving the accuracy of weather predictions, helping mitigate the impact of natural disasters. Quantum-enhanced environmental modeling is enabling more accurate simulations of climate change and its effects on ecosystems. Entanglement, a fundamental concept in quantum mechanics, plays a pivotal role in harnessing the power of quantum information. At the heart of entanglement is the idea that particles, when correlated, can exist in states where the properties of one particle are instantaneously connected to the properties of another, regardless of the distance between them. This phenomenon defies classical intuition and serves as a valuable resource in quantum information processing. Quantum teleportation, one of the remarkable applications of entanglement, allows for the transmission of an unknown quantum state from one location to another. The process relies on entangled particles to transmit quantum information accurately and securely, providing a foundation for quantum communication and quantum networking. Quantum cryptography, closely related to entanglement, leverages the principles of quantum mechanics to create secure communication channels that are theoretically immune to eavesdropping. Quantum key distribution (QKD) protocols, such as the BBM92 protocol developed by Charles Bennett and Gilles Brassard, establish secret cryptographic keys between two parties in a way that any interception of the key can be detected. This technology is being deployed in secure communication networks, financial institutions, and government agencies to protect sensitive information from potential quantum attacks. Quantum error correction is an essential aspect of harnessing entanglement for quantum information processing. Quantum systems are susceptible to errors and decoherence, which can disrupt quantum computations. Quantum error correction codes, such as the surface code, provide a means to detect and correct errors that may occur during quantum operations. Efforts in quantum error correction are vital for building practical and reliable quantum computers. Quantum algorithms, another key component of quantum information, utilize entanglement to perform certain calculations at speeds unattainable by classical computers. Shor's algorithm, for instance, can factor large numbers exponentially faster than classical algorithms, posing a significant challenge to classical encryption methods. Grover's algorithm, on the other hand, accelerates database searching and unstructured search problems by providing a quadratic speedup. These quantum algorithms are fundamental to quantum information theory and have applications across various domains, from cryptography to optimization and machine learning. Quantum computing, one of the most prominent applications of quantum information, leverages entangled qubits to perform calculations beyond the reach of classical computers. Quantum computers exploit the principles of quantum parallelism and entanglement to solve complex problems efficiently. While quantum hardware is still in its early stages of development, it holds the promise of revolutionizing fields such as materials science, drug discovery, and cryptography. Quantum communication networks, a practical application of quantum entanglement, are emerging as secure means of transmitting information. These networks rely on the principles of quantum key distribution (QKD) to establish secure communication channels that are resistant to quantum attacks. The development of quantum repeaters and quantum satellites is extending the reach of secure quantum communication over longer distances. Quantum-enhanced sensing technologies, such as quantum gravimeters and quantum magnetometers, take advantage of entanglement to achieve unprecedented levels of sensitivity and precision. These sensors have applications in geophysics, mineral exploration, and navigation, providing valuable insights and improving the accuracy of measurements. Quantum-enhanced imaging techniques, such as quantum-enhanced lidar and quantum-enhanced microscopy, offer advanced imaging capabilities for applications in remote sensing and biological research. These techniques provide greater accuracy, resolution, and depth perception, leading to breakthroughs in scientific discovery and industrial applications. Quantum-enhanced metrology, a growing field in quantum information theory, enhances the precision of measurements, including timekeeping and spectroscopy. Quantum clocks, which utilize the quantum properties of entangled particles, have the potential to redefine timekeeping standards with unprecedented accuracy. Quantum-enhanced cryptography is a critical area of research focused on developing cryptographic protocols that are secure against quantum attacks. Post-quantum cryptographic algorithms aim to protect sensitive information from potential threats posed by quantum computers. Quantum-resistant encryption schemes, digital signatures, and key exchange protocols are being designed to ensure the security of data in a post-quantum world. Quantum-enhanced algorithms in machine learning, a rapidly growing field, leverage the power of entangled qubits to improve data analysis, classification, and predictive modeling. These algorithms can process and analyze large datasets more efficiently, making them valuable in fields where data-driven decisions are crucial, such as finance, healthcare, and recommendation systems. Quantum-enhanced artificial intelligence is a frontier in quantum information theory that seeks to enhance the training and performance of machine learning models. Quantum algorithms have the potential to reduce the time required for training deep neural networks and improve optimization processes, leading to advancements in natural language processing and computer vision. Quantum-enhanced materials science is at the forefront of quantum information theory, revolutionizing the discovery and development of novel materials with exceptional properties. Quantum computers can simulate and analyze the behavior of molecules and materials at the quantum level, providing insights into materials for electronics, energy storage, and pharmaceuticals. Quantum-enhanced optimization, an application with wide-ranging impact, addresses complex optimization problems in fields such as logistics, finance, and supply chain management. Quantum algorithms like the Quantum Approximate Optimization Algorithm (QAOA) offer the potential to find optimal solutions more efficiently than classical algorithms. Quantum simulations, another application of quantum information theory, enable researchers to model and understand complex quantum systems, from molecules to materials. These simulations have applications in quantum chemistry, condensed matter physics, and quantum field theory, driving advancements in scientific understanding and technological innovation. Quantum-enhanced finance models are transforming risk assessment, portfolio optimization, and investment strategies in the financial industry. Quantum algorithms can perform complex financial simulations and provide more accurate insights into market trends and investment decisions. Quantum-enhanced logistics and supply chain management are optimizing resource allocation, transportation routes, and inventory management. This has implications for reducing costs, minimizing environmental impact, and improving the efficiency of global supply chains. Quantum-enhanced drug discovery is accelerating the development of new pharmaceuticals by simulating molecular interactions and identifying potential drug candidates more quickly and accurately. This has the potential to revolutionize healthcare by bringing life-saving drugs to market faster.

Chapter 3: Advanced Quantum Algorithms and Quantum Complexity

Exploring quantum algorithm complexity is a journey into the heart of quantum computing's computational power and limitations. In the realm of classical computing, algorithms are evaluated based on their time and space complexity, providing insights into their efficiency. Quantum algorithms, however, introduce a new layer of complexity that challenges our understanding of computation. One of the fundamental concepts in quantum algorithm complexity is the notion of quantum speedup. Quantum speedup refers to the advantage that quantum algorithms have over their classical counterparts in terms of computation time. Quantum algorithms can solve certain problems exponentially faster than the best-known classical algorithms. This exponential speedup arises from the principles of superposition and entanglement, which enable quantum computers to explore multiple solutions in parallel. Shor's algorithm, a prime example of quantum speedup, can factor large numbers exponentially faster than classical algorithms, posing a significant threat to classical encryption methods. The concept of quantum oracle complexity adds depth to the exploration of quantum algorithms. In quantum oracle complexity, an oracle represents an abstract black-box function that quantum algorithms use to query for information. The number of queries, or calls, to the oracle plays a crucial role in determining the algorithm's efficiency. Quantum algorithms often aim to minimize the number of oracle queries required to solve a specific problem, leading to improved computational efficiency. Quantum algorithm complexity extends beyond oracle-based models to encompass a broader range of computational problems. The study of quantum decision trees explores the complexity of decision problems that can be solved efficiently by quantum algorithms. Quantum decision trees provide insights into the number of queries required to make binary decisions, making them a valuable tool for understanding quantum algorithmic efficiency. Quantum query complexity is another facet of quantum algorithm complexity that focuses on the minimum number of queries required to solve a specific problem. This concept highlights the efficiency gains that quantum algorithms can achieve by minimizing the number of oracle queries. Quantum query complexity can be used to analyze algorithms for various tasks, from database searching to optimization. Grover's algorithm, a quantum search algorithm, is a prime example of quantum query complexity in action. Grover's algorithm offers a quadratic speedup over classical search algorithms, reducing the number of queries required to find a specific item in an unsorted database. Quantum algorithm complexity also encompasses the study of quantum communication complexity. In quantum communication complexity, two or more parties collaborate to solve a computational problem by exchanging quantum information. The goal is to minimize the amount of quantum communication required to achieve a specific task. Quantum communication complexity has applications in distributed computing, secure multi-party computation, and collaborative problem-solving. Quantum lower bounds represent an essential aspect of quantum algorithm complexity. These bounds define the limits of what can be achieved by quantum algorithms for specific problems. Quantum lower bounds help identify problems that are inherently hard for quantum computers, shedding light on the boundaries of quantum computational power. The study of quantum lower bounds involves proving that certain problems require a minimum number of quantum queries or operations to be solved, even by the most efficient quantum algorithms. Quantum complexity classes provide a framework for categorizing quantum algorithms based on their computational power. Quantum analogs of classical complexity classes, such as BQP (bounded-error quantum polynomial time) and QMA (quantum Merlin-Arthur), help classify problems that can be efficiently solved by quantum computers. These classes offer insights into the relationships between quantum and classical computational complexity and help identify problems where quantum algorithms excel. Quantum complexity classes also help define the boundaries between problems that are efficiently solvable by quantum computers and those that are not. Quantum algorithm complexity extends to the study of quantum communication protocols. Quantum protocols, such as quantum key distribution (QKD) and secure multi-party computation, leverage quantum entanglement and superposition to achieve secure and efficient communication. Quantum cryptography, a field closely related to quantum algorithm complexity, explores the security of quantum communication protocols. Quantum-resistant cryptography, on the other hand, focuses on developing cryptographic algorithms that remain secure in a post-quantum world. These algorithms are designed to withstand attacks from quantum computers, which have the potential to break classical encryption methods. Quantum algorithm complexity also plays a role in the study of quantum machine learning. Quantum machine learning algorithms leverage quantum computation to enhance data analysis, classification, and predictive modeling. The study of quantum machine learning complexity examines the efficiency gains and limitations of these algorithms compared to classical machine learning techniques. Quantum-enhanced optimization algorithms, another area of interest, aim to solve complex optimization problems more efficiently than classical counterparts. Quantum optimization algorithms, such as the Quantum Approximate Optimization Algorithm (QAOA), offer the potential for significant speedup in finding optimal solutions. Understanding the computational complexity of quantum optimization algorithms helps assess their practical utility in solving real-world optimization problems. Quantum algorithm complexity also intersects with the study of quantum simulation. Quantum computers have the potential to simulate quantum systems, offering insights into molecular interactions, materials science, and quantum chemistry. Quantum simulations provide a valuable tool for exploring the behavior of quantum systems and optimizing the design of new materials. Quantum algorithm complexity is a dynamic field that continues to evolve as researchers uncover new insights into the capabilities and limitations of quantum computation. As quantum hardware advances and quantum algorithms become more sophisticated, the exploration of quantum algorithm complexity will play a crucial role in shaping the future of quantum computing and its applications.

Advanced quantum algorithm design represents the pinnacle of harnessing the power of quantum computation. Building upon the foundational principles of quantum mechanics and algorithmic theory, advanced quantum algorithms push the boundaries of what is possible in terms of computational efficiency and problem-solving capabilities. These algorithms are designed to address complex problems that are beyond the reach of classical computers, offering exponential speedup and innovative solutions. One of the key challenges in advanced quantum algorithm design is finding ways to efficiently leverage quantum parallelism and entanglement to achieve practical computational advantages. This requires a deep understanding of quantum circuit design, quantum gate operations, and the quantum oracle model. Quantum circuits are at the heart of advanced quantum algorithms, representing the sequence of quantum gates and operations that manipulate qubits to perform calculations. Designing efficient quantum circuits often involves optimizing gate sequences to minimize the overall quantum computational resources required. Quantum gate operations, including the Hadamard gate, CNOT gate, and Toffoli gate, are the building blocks of quantum circuits.

Advanced quantum algorithm designers must select and combine these gates strategically to achieve desired computational outcomes. The quantum oracle model is a crucial concept in advanced quantum algorithms, allowing quantum algorithms to query for information in black-box functions. Efficiently utilizing quantum oracles is essential for many quantum algorithms, such as Grover's algorithm and quantum algorithms for solving optimization problems. Grover's algorithm, a seminal example of advanced quantum algorithm design, accelerates database searching and unstructured search problems. It offers a quadratic speedup over classical algorithms, reducing the number of queries required to find a specific item in an unsorted database. Quantum algorithms for solving optimization problems are another area of significant advancement in quantum algorithm design. These algorithms, like the Quantum Approximate Optimization Algorithm (QAOA), aim to find optimal solutions to complex optimization problems more efficiently than classical algorithms. QAOA leverages quantum parallelism and gate operations to explore the solution space and converge to near-optimal solutions. The design of advanced quantum algorithms often involves addressing computational challenges in various domains, including cryptography, machine learning, and materials science. In cryptography, advanced quantum algorithms explore the field of post-quantum cryptography, which aims to develop cryptographic methods that remain secure in the face of quantum attacks. Quantum-resistant encryption schemes, digital signatures, and key exchange protocols are designed to protect sensitive information in a post-quantum world. Machine learning is another domain where advanced quantum algorithms are making significant contributions. Quantum machine learning algorithms enhance data analysis, classification, and predictive modeling by exploiting quantum parallelism and gate operations. These algorithms have the potential to revolutionize fields where data-driven decisions are crucial, such as finance, healthcare, and recommendation systems. Quantum-enhanced artificial intelligence extends the capabilities of machine learning models by improving training processes and optimization tasks. Quantum algorithms can accelerate the training of deep neural networks and enhance the performance of natural language processing and computer vision models. Materials science is yet another domain where advanced quantum algorithms play a transformative role. Quantum computers can simulate and analyze the behavior of molecules and materials at the quantum level with unparalleled accuracy. This capability has applications in designing novel materials for electronics, optimizing drug molecules for pharmaceuticals, and understanding molecular interactions for chemistry and materials science. The design of advanced quantum algorithms also involves tackling challenges related to quantum error correction and fault tolerance. Quantum systems are inherently susceptible to errors and decoherence, which can disrupt quantum computations. Advanced quantum algorithm designers must work in conjunction with quantum hardware engineers to develop error-correcting codes and fault-tolerant techniques that ensure the reliability and stability of quantum computations. Quantum error correction codes, such as the surface code, are essential for detecting and correcting errors that may occur during quantum operations. Efforts in quantum error correction are fundamental for building practical and scalable quantum computers. Fault tolerance is another critical aspect of advanced quantum algorithm design. It involves developing algorithms that can maintain their performance even in the presence of errors and noise in the quantum hardware. Quantum fault-tolerant algorithms are designed to mitigate the impact of errors, ensuring the correctness of quantum computations. Quantum algorithm designers also explore the concept of quantum compilation, which involves translating high-level quantum algorithms into sequences of quantum gates and operations *that can be executed on a specific quantum computer. Quantum compilation is crucial for optimizing the efficiency and performance of quantum circuits on different quantum hardware platforms. Efficient quantum compilation tools and techniques are essential for practical quantum algorithm implementation. Advanced quantum algorithm design is an interdisciplinary field that requires collaboration between quantum physicists, computer scientists, mathematicians, and domain experts in various application areas. It involves theoretical analysis, algorithmic creativity, and experimental validation on quantum hardware. Quantum algorithm designers must navigate the delicate balance between the theoretical elegance of algorithms and their practical feasibility on existing or near-future quantum hardware. The field of quantum computing is continually evolving, with ongoing research into new quantum algorithms and their applications. Advanced quantum algorithm designers are at the forefront of this rapidly developing field, pushing the boundaries of quantum computational capabilities and addressing some of the most complex and challenging problems in science and technology. As quantum hardware continues to advance, the impact of advanced quantum algorithms on industries and scientific discovery is expected to grow exponentially, opening up new possibilities and revolutionizing the way we approach computational problems.*

Chapter 4: Quantum Machine Learning and Quantum Artificial Intelligence

Quantum machine learning represents a cutting-edge intersection of quantum computing and artificial intelligence, where quantum algorithms and techniques are applied to enhance machine learning models. Machine learning, a field focused on creating algorithms that enable computers to learn and make predictions from data, has seen remarkable progress in recent years. Quantum machine learning aims to leverage the unique properties of quantum computers, such as superposition and entanglement, to accelerate and improve various aspects of machine learning. One of the fundamental advantages of quantum machine learning is the potential for exponential speedup in solving specific problems compared to classical machine learning algorithms. Quantum algorithms, like quantum support vector machines (QSVM) and quantum principal component analysis (PCA), have demonstrated the ability to process and analyze data more efficiently. QSVM, for example, can classify data in a quantum-enhanced manner, offering significant computational advantages over classical support vector machines. Quantum machine learning also extends to unsupervised learning tasks, where quantum algorithms like quantum clustering and quantum singular value decomposition (SVD) aim to uncover patterns and structure in data. Quantum clustering algorithms can group data points into clusters more efficiently, while quantum SVD can extract essential information from large datasets. Quantum machine learning is not limited to classical data analysis but extends to quantum data as well. Quantum data can be inherently complex due to its entangled and superposed nature. Quantum machine learning algorithms, like quantum Boltzmann machines and quantum Hopfield networks, are designed to model and process quantum data effectively. These algorithms have applications in quantum chemistry, quantum materials science, and quantum information processing. Quantum-enhanced optimization techniques play a vital role in quantum machine learning. Quantum optimization algorithms, such as the Quantum Approximate Optimization Algorithm (QAOA), can find optimal solutions to complex problems more efficiently. QAOA leverages quantum parallelism to explore the solution space and converge to near-optimal solutions. Quantum machine learning models also offer the potential to accelerate training processes for classical machine learning models. Quantum-enhanced gradient descent algorithms, for example, can optimize the parameters of machine learning models more rapidly. These quantum machine learning models have applications in training deep neural networks and improving optimization tasks. Quantum machine learning algorithms can provide a valuable boost in solving combinatorial optimization problems. Tasks like the traveling salesman problem, where finding the shortest route through a set of cities is computationally challenging, can benefit from quantum algorithms that explore solution spaces efficiently. Quantum annealers, such as those developed by D-Wave Systems, offer a quantum-inspired approach to solving optimization problems and have found applications in various industries. Quantum machine learning can also enhance recommendation systems and personalized marketing. Quantum algorithms can process vast amounts of user data to provide tailored recommendations, leading to improved user experiences and increased sales. Quantum machine learning models have the potential to address challenges in financial modeling and risk assessment. These models can analyze complex financial data and optimize investment portfolios, reducing risks and maximizing returns. Quantum-enhanced machine learning is at the forefront of developments in natural language processing (NLP) and sentiment analysis. Quantum algorithms can process and understand textual data more efficiently, enabling applications in chatbots, sentiment analysis, and language translation. Quantum machine learning models can be applied to healthcare and drug discovery. Quantum algorithms can simulate molecular interactions, speeding up drug discovery processes and improving the development of pharmaceuticals. Quantum machine learning also plays a role in optimizing supply chains and logistics. These models can analyze data related to transportation, inventory management, and resource allocation, leading to cost reductions and improved efficiency. Quantum-enhanced machine learning is contributing to advancements in autonomous systems. Quantum algorithms can enhance the perception and decision-making capabilities of autonomous vehicles, drones, and robotics. Quantum machine learning has applications in cybersecurity, where quantum algorithms can improve threat detection and network security. Quantum-resistant cryptography, an emerging field, aims to protect data from potential quantum attacks by designing cryptographic algorithms that are secure in a post-quantum world. Quantum machine learning algorithms can also improve weather forecasting by analyzing complex meteorological data. These algorithms can enhance the accuracy of predictions and mitigate the impact of natural disasters. Quantum machine learning is poised to revolutionize environmental modeling. Quantum algorithms can simulate climate change scenarios and ecosystem dynamics, providing valuable insights for environmental conservation and policy. Quantum-enhanced machine *learning is driving advancements in energy optimization. These models can optimize energy grids, reduce consumption, and improve the efficiency of energy production and distribution. Quantum machine learning is a rapidly evolving field that continues to push the boundaries of what is possible in machine learning and artificial intelligence. Researchers and practitioners in quantum machine learning are at the forefront of developing innovative algorithms and techniques that have the potential to transform industries and address some of the most complex and pressing challenges in the modern world. As quantum hardware continues to advance, the impact of quantum machine learning on various domains is expected to grow, ushering in a new era of computational capabilities and data-driven insights. Quantum AI, a burgeoning field at the intersection of quantum computing and artificial intelligence, represents a profound shift in our approach to solving complex problems. While classical AI has made significant strides in recent years, quantum AI promises to take computation to new heights, transcending the limits of classical computing. The central idea behind quantum AI is harnessing the power of quantum mechanics to perform computations that are practically impossible for classical computers to achieve in a reasonable time frame. Quantum AI models and algorithms leverage quantum properties such as superposition, entanglement, and quantum parallelism to process and analyze data with unprecedented speed and efficiency. One of the key areas where quantum AI stands to make a transformative impact is in optimization problems. Optimization problems are ubiquitous in science, engineering, logistics, and many other fields. Classical optimization algorithms face limitations when dealing with large-scale, combinatorial, or non-linear problems. Quantum optimization algorithms, like the*

Quantum Approximate Optimization Algorithm (QAOA), have shown remarkable promise in finding near-optimal solutions to complex optimization problems exponentially faster than their classical counterparts. These quantum algorithms have practical applications in supply chain management, portfolio optimization, and resource allocation. Quantum AI also holds great potential in the realm of machine learning. Classical machine learning algorithms have made significant progress in tasks like image recognition, natural language processing, and recommendation systems. However, as datasets continue to grow in size and complexity, classical algorithms face challenges in training and inference. Quantum machine learning models aim to address these challenges by speeding up the training process and improving the accuracy of models. Quantum algorithms like the Quantum Support Vector Machine (QSVM) and quantum neural networks have the potential to revolutionize the field by providing exponential speedup in certain machine learning tasks. QSVM, for instance, can classify data points in a quantum-enhanced manner, offering computational advantages over classical support vector machines. Quantum neural networks leverage quantum parallelism to accelerate training and optimization processes, making them suitable for deep learning tasks. Moreover, quantum AI can enhance the capabilities of reinforcement learning algorithms, enabling autonomous systems like self-driving cars and robotics to make faster and more accurate decisions. Another realm where quantum AI promises breakthroughs is in the field of quantum chemistry and drug discovery. Simulating molecular interactions and understanding the behavior of complex molecules is a computationally intensive task. Quantum computers can simulate quantum systems with unrivaled accuracy, making them ideal for modeling molecular structures and predicting chemical reactions. This capability has significant implications for pharmaceutical research, as quantum AI can expedite drug discovery by identifying potential drug candidates and optimizing molecular structures more efficiently. Quantum AI can also contribute to materials science by simulating and analyzing the properties of novel materials at the quantum level. This can lead to the development of advanced materials with unique properties, such as superconductors, catalysts, and materials for energy storage. Additionally, quantum AI can enhance computational chemistry methods by providing more accurate simulations and predictions for chemical processes. Quantum AI is poised to revolutionize the field of cryptography. As quantum computers become more powerful, they pose a significant threat to classical encryption methods. Quantum-resistant cryptographic algorithms are being developed to secure data in a post-quantum world. These cryptographic techniques aim to protect sensitive information from potential quantum attacks. Quantum key distribution (QKD) protocols, such as the BBM92 protocol, are designed to establish secure communication channels that are theoretically immune to eavesdropping by quantum computers. Quantum AI models and algorithms can also play a crucial role in enhancing cybersecurity. They can improve threat detection, intrusion detection, and anomaly detection by analyzing network traffic and identifying suspicious patterns. This can lead to more robust cybersecurity measures to safeguard critical infrastructure and sensitive data. Quantum AI has the potential to transform the financial industry by optimizing trading strategies, risk assessment, and portfolio management. Quantum algorithms can analyze vast amounts of financial data, leading to more accurate predictions and improved decision-making. Quantum AI models can contribute to climate modeling and *environmental science by simulating climate change scenarios and predicting the impact of various factors on ecosystems. These simulations can provide valuable insights for conservation efforts and environmental policy. Quantum AI can also enhance the field of quantum metrology by improving the precision of measurements, including timekeeping, spectroscopy, and gravitational wave detection. Quantum-enhanced sensors and detectors can revolutionize various scientific and industrial applications. Quantum AI models have applications in natural language processing (NLP) by processing and understanding textual data more efficiently. This can lead to advances in chatbots, sentiment analysis, and language translation, improving communication and information retrieval. Quantum AI can advance autonomous systems by enhancing the perception and decision-making capabilities of autonomous vehicles, drones, and robotics. Quantum algorithms can process sensor data and make real-time decisions, leading to safer and more efficient autonomous systems. Quantum AI is a rapidly evolving field with the potential to revolutionize industries, scientific research, and everyday life. Researchers and practitioners in quantum AI are pushing the boundaries of what is possible in computation, optimization, and machine learning. As quantum hardware continues to advance, the impact of quantum AI on various domains is expected to grow, ushering in a new era of computational capabilities and data-driven insights. Quantum AI represents a paradigm shift in how we approach complex problems, offering unprecedented computational power and innovative solutions to some of the most challenging issues facing humanity.*

Chapter 5: Quantum Simulation at the Cutting Edge

Advanced quantum simulation methods represent a cornerstone of quantum computing, offering the ability to simulate complex quantum systems with unparalleled accuracy and efficiency. Quantum simulation harnesses the intrinsic properties of quantum mechanics to model the behavior of quantum systems, such as molecules, materials, and physical phenomena. These simulations enable researchers to gain insights into the quantum world and solve problems that are computationally intractable for classical computers. Quantum simulations leverage the concept of qubits, the fundamental units of quantum information, to represent and manipulate quantum states. One of the primary applications of quantum simulations is in quantum chemistry, where researchers use quantum computers to study the electronic structure of molecules and predict chemical reactions. Quantum algorithms, like the variational quantum eigensolver (VQE) and the quantum phase estimation algorithm, are employed to solve the electronic structure problem efficiently. The electronic structure problem involves determining the ground-state energy and properties of molecules, which are crucial for drug discovery, materials science, and understanding chemical reactions. Quantum simulations in chemistry can provide insights into the behavior of complex molecules and help design new materials with specific properties, such as superconductors or catalysts. Another area where advanced quantum simulation methods shine is in quantum materials science, where researchers investigate the properties of novel materials with quantum effects. Quantum computers can simulate the behavior of electrons in materials, leading to the discovery of materials with unique properties or the optimization of existing ones. For example, quantum simulations can uncover high-

temperature superconductors, materials with zero electrical resistance at elevated temperatures, which have significant implications for energy transmission and storage. Advanced quantum simulation methods can be used to study quantum phase transitions, a fundamental concept in condensed matter physics. These simulations help researchers understand how materials transition between different quantum states and gain insights into exotic phenomena like topological insulators and quantum spin liquids. Quantum simulations also play a crucial role in quantum field theory, a framework for describing the fundamental forces of nature. Quantum field theory simulations aim to understand phenomena at the quantum level, such as particle interactions and the behavior of matter and energy in extreme conditions, like those found in the early universe. Advanced quantum simulation methods can facilitate research in quantum field theory, potentially leading to breakthroughs in our understanding of the fundamental forces that govern the universe. Quantum simulations can also impact the field of quantum computing itself. Quantum error correction is a critical component of building practical and scalable quantum computers. Quantum simulations can be used to model the behavior of quantum error correction codes and optimize their performance. This allows researchers to design more efficient codes that mitigate the effects of noise and decoherence in quantum hardware. In the realm of quantum simulations, researchers are exploring the concept of quantum approximate optimization algorithms (QAOA). QAOA algorithms are designed to solve optimization problems by preparing quantum states that approximate the optimal solution. These algorithms leverage variational techniques to find near-optimal solutions efficiently, making them valuable in addressing complex optimization problems across various domains. Quantum simulations are not limited to closed quantum systems but also extend to open quantum systems. Open quantum systems interact with their environment, leading to phenomena like decoherence and dissipation. Quantum simulations of open systems help researchers understand and control the effects of decoherence, which is crucial for maintaining the coherence of quantum states in practical quantum computing applications. Advanced quantum simulation methods have practical applications in the development of quantum algorithms and quantum machine learning. Simulating quantum circuits and quantum algorithms on quantum hardware can provide insights into their performance and potential improvements. Quantum simulations enable researchers to test new algorithms and assess their scalability and efficiency. Additionally, quantum simulations can assist in optimizing quantum machine learning models by exploring their behavior in complex quantum environments. Quantum machine learning algorithms can benefit from quantum simulations by leveraging quantum parallelism and gate operations to accelerate training and inference processes. Quantum simulations can also be used to analyze the behavior of quantum neural networks and quantum-enhanced machine learning models. Quantum-enhanced optimization algorithms, like the Quantum Approximate Optimization Algorithm (QAOA), leverage quantum simulations to find optimal solutions to complex optimization problems more efficiently. QAOA explores the solution space by varying the parameters of a quantum circuit and converging to near-optimal solutions. Quantum simulations are instrumental in testing and optimizing these quantum optimization algorithms. Quantum simulations can also be applied to quantum cryptography, a field focused on securing communication using the principles of quantum mechanics. Quantum key distribution (QKD) protocols, such as the BBM92 protocol and the E91 protocol, rely on quantum simulations to assess the security and performance of quantum key exchange processes. These simulations help identify potential vulnerabilities and ensure the resilience of quantum cryptographic systems against eavesdropping attacks.

Quantum simulations offer researchers a powerful tool for exploring the behavior of quantum systems and solving complex problems across various domains. As quantum hardware continues to advance, the capabilities of quantum simulations will expand, opening new frontiers in scientific discovery, materials design, optimization, and quantum technology development. Quantum simulations are poised to play a pivotal role in advancing our understanding of the quantum world and driving innovation across a wide range of fields.

Quantum simulation stands as a remarkable avenue for tackling the profound complexities of quantum and classical systems. Its power lies in its ability to use quantum computers to emulate the behavior of these intricate systems, thereby unlocking insights and solutions that would be beyond the reach of classical computers. The promise of quantum simulation extends across a diverse array of disciplines, from fundamental quantum physics to condensed matter physics, materials science, and beyond. The fundamental concept underpinning quantum simulation is the capacity of quantum computers to manipulate quantum bits, or qubits, in ways that classical bits simply cannot replicate. Through the application of quantum gates and quantum algorithms, researchers can encode, process, and analyze the quantum states of complex systems. One of the foundational areas where quantum simulation holds immense potential is quantum chemistry. In quantum chemistry, understanding the quantum mechanical behavior of atoms and molecules is central to discovering new materials, designing drugs, and comprehending chemical reactions. Quantum computers, armed with quantum algorithms such as the Variational Quantum Eigensolver (VQE) and the Quantum Phase Estimation algorithm, have the capability to solve the electronic structure problem efficiently. This problem involves calculating the ground-state energy and properties of molecules, which are pivotal in fields like drug discovery and materials science. Quantum chemistry simulations can yield insights into the behavior of intricate molecules and facilitate the development of novel materials with tailored properties. Quantum simulation methods also extend their reach to the field of quantum materials science, where researchers explore materials with emergent quantum phenomena. Quantum computers can simulate the quantum behavior of electrons within materials, offering the potential to identify materials with exotic properties or optimize existing ones. For instance, these simulations have led to the discovery of high-temperature superconductors, materials with zero electrical resistance at elevated temperatures, which have profound implications for energy transmission and storage. Furthermore, quantum simulations delve into the realm of quantum phase transitions, where materials undergo quantum-driven changes in their properties. Studying quantum phase transitions aids in uncovering topological insulators and quantum spin liquids, both of which possess unusual quantum properties. Quantum simulations also make a significant impact in the realm of quantum field theory, a framework for explaining the fundamental forces governing the universe. Simulating quantum field theories allows researchers to investigate particle interactions and the behavior of matter and energy under extreme conditions. These simulations have the potential to provide profound insights into the building blocks of the universe. Advanced quantum simulation methods extend their influence into the domain of quantum error correction. Quantum computers are inherently susceptible to errors and decoherence, making error correction vital for building practical and reliable quantum computers. Quantum simulations can model the behavior of quantum error correction codes and optimize their performance. This contributes to the development of more efficient codes that mitigate the effects of noise and decoherence in quantum hardware. Quantum error correction codes, such as the surface code, are essential for *identifying and rectifying errors that may occur during quantum operations. In conjunction with quantum error correction, advanced quantum simulation techniques explore the concept of quantum approximate optimization algorithms (QAOA). QAOA algorithms aim to efficiently solve optimization problems by preparing quantum states that closely approximate the optimal solution. These algorithms employ variational techniques to find near-optimal solutions, making them invaluable for addressing complex optimization problems across various domains. Quantum simulations offer more than just insights into quantum phenomena; they also have practical applications in developing quantum algorithms and quantum machine learning. Simulating quantum circuits and quantum algorithms on quantum hardware provides valuable insights into their performance and potential enhancements. These simulations enable researchers to test new algorithms, assess their scalability and efficiency, and identify areas for improvement. Moreover, quantum simulations support the optimization of quantum machine learning models by exploring their behavior in complex quantum environments. Quantum machine learning algorithms can benefit from the quantum parallelism and gate operations provided by quantum simulations, accelerating both training and inference processes. Quantum simulations can help analyze the behavior of quantum neural networks and quantum-enhanced machine learning models. Advanced quantum simulation methods have direct implications for quantum cryptography, a field dedicated to securing communications using the principles of quantum mechanics. The advent of powerful quantum computers poses a significant threat to classical encryption methods. Quantum-resistant cryptographic algorithms are being developed to safeguard data in a post-quantum era. These cryptographic techniques aim to protect sensitive information from potential quantum attacks. Quantum key distribution (QKD) protocols, such as the BBM92 protocol and the E91 protocol, rely on quantum simulations to assess the security and performance of quantum key exchange processes. These simulations play a vital role in identifying potential vulnerabilities and ensuring the robustness of quantum cryptographic systems against eavesdropping attacks. Quantum simulations also have the potential to transform cybersecurity. By analyzing network traffic and detecting unusual patterns, quantum AI models can enhance threat detection, intrusion detection, and anomaly detection. This leads to more robust cybersecurity measures, protecting critical infrastructure and sensitive data. Quantum simulations extend their influence into financial modeling and risk assessment. They can analyze vast amounts of financial data to provide accurate predictions and enhance decision-making in areas like trading strategies, portfolio management, and risk assessment. Quantum-enhanced machine learning models can contribute to the development of autonomous systems. By processing sensor data and making real-time decisions, quantum AI models can enhance the perception and decision-making capabilities of autonomous vehicles, drones, and robotics. In summary, quantum simulation methods are at the forefront of scientific research, technological innovation, and problem-solving across diverse fields. Their ability to emulate and analyze complex quantum and classical systems has the potential to revolutionize our understanding of the quantum world and our approach to addressing some of the most pressing challenges in science, technology, and society.*

Chapter 6: Quantum Error Correction: Beyond the Basics

Advanced quantum error correction schemes represent a pivotal aspect of the ongoing quest to build robust and scalable quantum computers. These schemes are indispensable in addressing the inherent fragility of quantum bits, or qubits, due to their susceptibility to noise, decoherence, and other quantum errors. Quantum error correction forms the bedrock of quantum computing, ensuring the reliability of quantum computations by detecting and rectifying errors that occur during quantum operations. One of the fundamental challenges in quantum error correction is preserving the fragile quantum information encoded in qubits while safeguarding against errors. To tackle this challenge, researchers have developed a wide array of quantum error correction codes and schemes that can protect quantum information from the ravages of noise. At the core of quantum error correction lies the concept of quantum error-correcting codes, which are analogous to classical error-correcting codes but adapted to the peculiarities of quantum information. These codes encode quantum states into a larger quantum space, introducing redundancy that allows for the detection and correction of errors. One of the earliest and most renowned quantum error-correcting codes is the three-qubit bit-flip code, which guards against errors that flip individual qubits from 0 to 1 or vice versa. The three-qubit bit-flip code employs an ingenious encoding scheme that enables the identification and rectification of single-qubit errors. However, as quantum computers grow in complexity and qubit counts increase, advanced quantum error correction schemes become imperative to address more intricate errors.

To comprehend the significance of advanced quantum error correction schemes, it is essential to recognize the nature of quantum errors and their origins. Quantum errors can arise from a multitude of sources, including fluctuations in the electromagnetic environment, imprecise gate operations, and thermal effects. These errors manifest as undesired rotations, phase shifts, or bit flips in the quantum states of qubits. Quantum error correction operates on the premise that errors are probabilistic and occur randomly during quantum computations. As a result, quantum error correction codes are designed to detect errors through the careful arrangement of qubits in encoded states. One of the pivotal concepts in quantum error correction is the quantum syndrome, a set of measurements that reveals information about the errors that have occurred. To identify and correct errors, quantum error correction codes employ syndrome measurements, which are akin to the "check bits" in classical error-correcting codes. The syndrome measurements are performed without directly measuring the quantum states of qubits, as doing so would collapse their superposition and render error correction ineffective. Instead, quantum error correction relies on the results of syndrome measurements to infer the presence and nature of errors. The process of error correction, often referred to as error syndromes and error correction cycles, iteratively detects and corrects errors to ensure the fidelity of quantum computations. Quantum error correction codes encompass a range of possibilities beyond the three-qubit bit-flip code. The five-qubit code, for instance, provides protection against both bit-flip and phase-flip errors, extending the error-correcting capabilities of quantum codes. The seven-qubit Steane code offers even more robust protection, addressing both types of errors and providing a path toward fault-tolerant quantum computing. However, to achieve fault tolerance, advanced quantum error correction schemes are essential. Fault tolerance is a crucial milestone in quantum computing, signifying the ability to perform reliable quantum computations despite the presence of errors. To attain fault tolerance, quantum error correction must not only detect and correct errors but also do so in a manner that does not propagate errors or degrade the quantum information. One of the pioneering concepts in fault-tolerant quantum error correction is the use of concatenated codes. Concatenated codes involve encoding qubits multiple times in a hierarchical fashion, creating a multi-layered structure of error correction. Each layer of encoding provides progressively higher levels of protection against errors, culminating in fault tolerance. Concatenated codes represent a crucial step toward the realization of practical and scalable quantum computers. Another fundamental element in advanced quantum error correction schemes is the use of surface codes. Surface codes are two-dimensional arrangements of qubits on a lattice, where qubits are encoded in a grid-like fashion. Surface codes offer several advantages, including efficient syndrome measurements and the ability to detect and correct multiple errors simultaneously. These properties make surface codes a promising candidate for achieving fault tolerance in quantum computing. Surface codes also exhibit a natural form of error correction, known as topological error correction. In topological error correction, errors are viewed as topological defects in the lattice of qubits. These defects can be identified and corrected without the need for explicit syndrome measurements, simplifying the error correction process. Topological error correction has been a focal point of research in the quest for fault-tolerant quantum computing. One of the noteworthy aspects of advanced quantum error correction schemes is their compatibility with quantum gates and operations. Quantum gates are the building blocks of quantum computations, and ensuring their compatibility with error correction is essential for practical quantum computing. Transversality, a property that allows quantum gates to be applied to encoded qubits without introducing additional errors, is a sought-after feature in advanced quantum error correction schemes. Transversality ensures that error correction remains effective even as quantum computations become more complex. Implementing fault-tolerant quantum error correction requires a combination of techniques, including the use of concatenated codes, surface codes, and topological error correction, along with the careful design of quantum gates that preserve the encoded information. Achieving fault tolerance in quantum computing is an ongoing endeavor, with researchers working tirelessly to develop advanced quantum error correction schemes that can withstand the challenges posed by noise, decoherence, and quantum errors. These schemes hold the promise of revolutionizing quantum computing by enabling practical, scalable, and error-resistant quantum computers that can tackle complex problems and accelerate scientific discovery across a multitude of disciplines. Quantum fault tolerance represents a critical milestone in the quest to build practical and scalable quantum computers that can reliably perform complex computations. Fault tolerance is a fundamental requirement for quantum computing, as quantum bits, or qubits, are highly susceptible to errors caused by environmental factors, gate imperfections, and decoherence. The ability to detect and correct errors during quantum computations is paramount for ensuring the accuracy and integrity of quantum algorithms. To appreciate the significance of quantum fault tolerance, one must understand the unique challenges posed by quantum errors and the need for robust error-correcting codes. Quantum errors can manifest as bit flips, phase flips, or undesired rotations in the quantum states of qubits. These errors occur probabilistically, and their correction necessitates the use of quantum error-correcting codes. Quantum error correction operates on the principle of encoding quantum information into larger quantum spaces, introducing redundancy that enables the detection and rectification of errors. One of the foundational quantum error-correcting codes is the three-qubit bit-flip code, designed to correct single-qubit errors. This code employs an ingenious encoding scheme that enables the identification and rectification of errors, laying the groundwork for more advanced quantum error correction codes. As quantum computers evolve, advanced quantum error correction schemes become imperative to address increasingly complex errors and protect against the effects of noise. To achieve fault tolerance, quantum error correction must not only detect and correct errors but also do so in a manner that does not propagate errors or degrade the quantum information. Concatenated codes represent a pivotal concept in fault-tolerant quantum error correction. These codes involve encoding qubits multiple times in a hierarchical fashion, creating layers of error correction. Each layer provides progressively higher levels of protection against errors, culminating in fault tolerance. Concatenated codes are instrumental in the pursuit of scalable and reliable quantum computers. Another crucial aspect of fault-tolerant quantum error correction is the use of surface codes. Surface codes are two-dimensional arrays of qubits arranged in a lattice, with qubits encoded in a grid-like fashion. These codes offer several advantages, including efficient syndrome measurements and the ability to detect and correct multiple errors simultaneously. Surface codes are a promising candidate for achieving fault tolerance in quantum computing due to their robust error correction properties. Furthermore, surface codes exhibit a form of natural error correction known as topological error correction. In topological error correction, errors are viewed as topological defects in the lattice of qubits. These defects can be identified and rectified without the need for explicit syndrome measurements, simplifying the error correction process. Topological error correction has emerged as a central focus of research in the pursuit of fault-tolerant quantum computing. Achieving fault tolerance also involves ensuring the compatibility of quantum gates and operations with error correction. Quantum gates are the fundamental building blocks of quantum computations, and their interaction with error correction is critical. Transversality, a property that allows quantum gates to be applied to encoded qubits without introducing additional errors, is highly desirable in fault-tolerant quantum error correction schemes. Transversality ensures that error correction remains effective even as quantum computations become more intricate. Implementing fault-tolerant quantum error correction requires a combination of techniques, including concatenated codes, surface codes, topological error correction, and gate design that preserves encoded information. Fault tolerance is an ongoing endeavor in the field of quantum computing, with researchers dedicated to developing advanced quantum error correction schemes capable of withstanding the challenges posed by quantum errors, noise, and decoherence. The ultimate goal of fault-tolerant quantum computing is to create practical and scalable quantum computers that can tackle complex problems across various disciplines. Quantum fault tolerance extends its reach beyond quantum error correction and encompasses the broader context of fault-tolerant quantum computation. In fault-tolerant quantum computation, the *objective is to perform reliable quantum computations despite the presence of errors. The achievement of fault tolerance implies that quantum computers can execute quantum algorithms accurately, even when exposed to noise and imperfections. This level of reliability is indispensable for quantum computers to become valuable tools in scientific research, engineering, and various applications. One of the fundamental principles of fault-tolerant quantum computation is the threshold theorem, which establishes the conditions under which fault tolerance can be achieved. The threshold theorem states that if the error rate for quantum gates and measurements is below a certain threshold, fault-tolerant quantum computation becomes feasible. This threshold is typically quite low, implying that quantum hardware must exhibit a high degree of fidelity to enable fault tolerance. The threshold theorem provides a rigorous foundation for the pursuit of fault-tolerant quantum computation. To achieve fault tolerance, quantum error correction must be integrated into the fabric of quantum algorithms and computations. This integration involves the use of quantum error correction codes, such as surface codes and concatenated codes, to protect quantum information and correct errors during quantum operations. Quantum fault tolerance also entails the implementation of error-correcting procedures, which continuously monitor the quantum state and apply error correction as needed. These procedures, often referred to as error syndromes and error correction cycles, are essential for maintaining the integrity of quantum computations. Furthermore, fault-tolerant quantum computation relies on the concept of logical qubits, which are encoded quantum bits that are robustly protected against errors. Logical qubits are composed of multiple physical qubits, allowing them to withstand the effects of noise and decoherence. These logical qubits serve as the foundation for fault-tolerant quantum algorithms and computations. The achievement of fault tolerance in quantum computation has far-reaching implications for various fields and applications. Quantum computers have the potential to solve complex problems in areas such as cryptography, materials science, optimization, and drug discovery. Fault tolerance ensures the reliability of quantum algorithms in these domains, enabling quantum computers to provide accurate and trustworthy results. Moreover, fault-tolerant quantum computation is essential for the realization of quantum supremacy, a milestone where quantum computers can outperform classical computers in specific tasks. Quantum supremacy opens the door to new frontiers in scientific research and technological innovation. Quantum fault tolerance represents a critical bridge between the promise of quantum computing and its practical implementation. As researchers continue to advance the field of quantum error correction and fault-tolerant quantum computation, the vision of practical, scalable, and reliable quantum computers comes closer to reality, ushering in a new era of computation and discovery.*

Chapter 7: Quantum Hardware Innovations and Quantum Processors

Innovations in quantum processor design are at the forefront of quantum computing research, pushing the boundaries of what is possible in terms of computational power and quantum advantage. Quantum processors are the heart of quantum computers, responsible for executing quantum algorithms and performing quantum computations. To fully appreciate the innovations in quantum processor design, it's essential to understand the unique principles that underpin quantum computing. At the core of quantum computing are qubits, quantum bits that can exist in superposition, representing both 0 and 1 simultaneously. This property of superposition enables quantum processors to perform complex calculations in parallel, making them potentially exponentially faster than classical processors for certain problems. However, qubits are inherently fragile and susceptible to errors, making their reliable manipulation a significant challenge in quantum processor design. One of the pioneering innovations in quantum processor design is the use of physical qubits, which can be implemented using various quantum technologies. These physical qubits can be realized using different physical systems, such as superconducting circuits, trapped ions, or topological qubits. Each of these quantum technologies has its unique advantages and challenges, driving innovation in quantum processor design. Superconducting qubits are among the most widely used in quantum processors, as they are relatively easy to manufacture and manipulate. These qubits rely on the quantum properties of superconducting circuits, which can carry electrical currents with zero resistance.

Innovations in superconducting qubit design have led to improved coherence times and reduced error rates, making them a prominent choice in quantum processors. Trapped ion qubits are another promising technology, where individual ions are trapped and manipulated using electromagnetic fields. Innovations in trapped ion qubit design have resulted in exceptionally long coherence times, allowing for precise quantum operations and error correction. Topological qubits, a more recent innovation, are qubits that leverage exotic states of matter to provide inherent error protection. These qubits are inherently fault-tolerant, making them a potential game-changer in quantum processor design. In addition to qubit technologies, innovations in quantum processor design involve optimizing the connectivity and architecture of qubits within the processor. Quantum processors are designed to perform quantum gates, which are akin to logical operations in classical computing. These gates manipulate the quantum states of qubits to perform calculations. Innovations in gate design and control mechanisms have led to the development of more versatile and efficient quantum processors. For example, the use of parametric gates and two-qubit gates has expanded the range of quantum algorithms that can be executed on quantum processors. Moreover, innovations in quantum processor architecture have focused on improving qubit connectivity, allowing for more efficient quantum computations. Enhanced connectivity enables qubits to interact with neighboring qubits, facilitating the execution of quantum algorithms and reducing the time required for quantum computations. Quantum processor design also encompasses the development of quantum error correction techniques. As mentioned earlier, qubits are susceptible to errors caused by environmental factors and noise. Innovations in error correction codes and quantum error correction schemes are essential for building reliable and *fault-tolerant quantum processors. Quantum processors must incorporate error correction codes like the surface code or the Steane code to protect against errors and preserve the integrity of quantum information. Furthermore, innovations in quantum processor design involve addressing the challenges posed by decoherence, which is the loss of quantum coherence over time. Decoherence limits the duration of quantum computations and impacts the reliability of quantum processors. Innovative techniques, such as dynamical decoupling and quantum error correction, aim to mitigate the effects of decoherence and extend the coherence times of qubits. Quantum processor design also encompasses the development of quantum gates and quantum algorithms that are specifically tailored to take advantage of quantum parallelism. Quantum gates, such as the Hadamard gate and the CNOT gate, are essential building blocks in quantum algorithms. Innovations in gate design aim to improve gate fidelities and reduce gate errors, enabling the execution of more complex quantum algorithms with higher accuracy. Quantum algorithms, such as Shor's algorithm and Grover's algorithm, leverage the unique properties of quantum processors to solve problems that are intractable for classical computers. Innovations in quantum algorithm design have led to the discovery of new quantum algorithms and applications across various domains, from cryptography to optimization. Quantum processor design also extends its reach into the realm of quantum networking and communication. Quantum processors can serve as nodes in quantum networks, facilitating secure quantum communication and the distribution of entangled qubits. Innovations in quantum processor design are instrumental in developing quantum repeaters and quantum routers that enable long-distance quantum communication. These innovations are paving the way for the realization of a quantum internet, where quantum information can be transmitted globally with unprecedented security. Moreover, innovations in quantum processor design are intertwined with advancements in quantum hardware and quantum software. Quantum hardware includes the physical components of quantum processors, such as qubits, gates, and control electronics. Innovations in quantum hardware involve improving the performance and reliability of these components, ultimately enhancing the overall capabilities of quantum processors. Quantum software encompasses the algorithms, compilers, and programming languages used to program and control quantum processors. Innovations in quantum software aim to simplify the programming of quantum processors and make them more accessible to a broader range of users. Quantum cloud computing platforms and quantum-as-a-service offerings are emerging as a result of these innovations, democratizing access to quantum processors and quantum computing resources. The field of quantum computing is continuously evolving, driven by innovations in quantum processor design, quantum hardware, quantum software, and quantum algorithms. These innovations are expanding the horizons of what is achievable with quantum computing and hold the potential to revolutionize industries, solve complex problems, and drive scientific discovery. As researchers and engineers push the boundaries of quantum processor design, we can anticipate even more remarkable breakthroughs and applications that will shape the future of quantum computing. Quantum hardware tailored for specialized applications is a burgeoning field within the realm of quantum technology, where quantum devices are engineered to address specific challenges and tasks. The versatility of quantum hardware lies in its ability to harness the unique properties of quantum mechanics to tackle problems that are difficult or impossible for classical computers. In the pursuit of specialized quantum hardware, researchers and engineers are exploring various quantum platforms, each with its strengths and suitability for particular applications. One of the leading quantum hardware platforms is superconducting qubits, which are highly tunable and can be manipulated with precision, making them versatile for a range of applications. Superconducting qubits have found utility in quantum annealing, a specialized quantum computation technique aimed at solving optimization problems. Quantum annealers employ quantum fluctuations to find the optimal solution to a given problem by exploring a complex landscape of possibilities. These specialized quantum devices are used in diverse fields, from financial modeling to drug discovery and supply chain optimization. Another prominent quantum hardware platform is trapped ion quantum processors, which excel in providing long qubit coherence times and high-fidelity quantum gates. These attributes make trapped ion devices suitable for quantum simulations, quantum chemistry, and quantum cryptography. In quantum simulations, trapped ion devices emulate the behavior of quantum systems, enabling the study of complex physical and chemical processes. Quantum chemistry simulations leverage trapped ion processors to understand molecular interactions, design new materials, and advance drug discovery. Quantum cryptography, with its emphasis on secure communication, benefits from the long coherence times of trapped ion qubits to ensure the integrity and confidentiality of quantum keys. Specialized quantum hardware also includes photonic quantum processors, which manipulate quantum information encoded in photons, the fundamental particles of light. These processors have a distinct advantage in quantum communication and quantum cryptography applications. Photonic quantum processors* facilitate the distribution of entangled photons over long distances, making them a cornerstone of quantum key distribution (QKD) systems. QKD protocols, like the BB84 protocol and the E91 protocol, rely on the properties of entangled photons to establish secure communication channels that are resistant to eavesdropping. Additionally, photonic quantum processors are integral to quantum teleportation experiments, where quantum states are transferred instantaneously from one location to another using entangled photons. Another innovative avenue in specialized quantum hardware involves quantum-enhanced sensors. Quantum sensors exploit the precision of quantum devices to measure physical quantities with exceptional accuracy. Specialized quantum hardware can be engineered to detect minute changes in various parameters, from magnetic fields and gravitational forces to temperature and pressure. These quantum sensors have applications in geophysics, environmental monitoring, navigation systems, and fundamental physics experiments. For example, quantum gravimeters, which use atom interferometry, can measure gravitational accelerations with remarkable precision. These sensors are employed in geophysical exploration to detect subsurface structures and variations in gravitational fields. Quantum-enhanced sensors also play a role in improving the accuracy of global positioning systems (GPS) and inertial navigation systems used in autonomous vehicles and aerospace applications. Specialized quantum hardware extends its reach into the realm of quantum-enhanced imaging and microscopy. Quantum imaging techniques leverage the quantum properties of entangled photons to surpass classical limits in imaging resolution and sensitivity. Quantum-enhanced imaging has applications in medical imaging, materials science, and remote sensing. For example, quantum-enhanced imaging systems can improve *the resolution of optical coherence tomography (OCT) scans, aiding in the early detection of diseases and enhancing medical diagnostics. In materials science, quantum-enhanced microscopy can reveal the atomic-scale structures of materials, facilitating the development of advanced materials with tailored properties. Specialized quantum hardware is also making strides in quantum-enhanced machine learning. Quantum machine learning models leverage quantum processors to accelerate the training and inference processes of machine learning algorithms. These models can optimize complex functions and solve optimization problems more efficiently than classical computers. Quantum-enhanced machine learning has applications in diverse domains, including finance, logistics, and drug discovery. In financial modeling, quantum machine learning algorithms can enhance portfolio optimization, risk assessment, and algorithmic trading strategies. Logistics and supply chain management benefit from quantum algorithms that optimize route planning and resource allocation. In drug discovery, quantum machine learning accelerates the identification of potential drug candidates and their interactions with biological systems. Quantum hardware for specialized applications is closely intertwined with quantum software development. The programming and control of quantum processors are essential components of leveraging their capabilities for specific tasks. Quantum software includes quantum algorithms, compilers, and programming languages that enable the efficient execution of quantum computations on specialized hardware. Quantum cloud computing platforms are emerging to provide access to quantum hardware resources and simplify the development and deployment of quantum applications. These platforms democratize access to specialized quantum hardware, making it available to a broader range of researchers, developers, and organizations. Moreover, specialized quantum hardware is at the forefront of quantum cryptography and quantum-safe encryption. Quantum cryptography protocols, such as QKD, rely on the secure distribution of quantum keys generated by quantum devices. Specialized quantum hardware ensures the reliability and security of quantum key distribution systems, safeguarding sensitive communications. Furthermore, quantum-safe encryption algorithms are being developed to protect data against potential attacks by quantum computers. These encryption schemes are designed to withstand quantum attacks and provide long-term security for digital communication. In summary, specialized quantum hardware represents a realm of innovation and exploration in quantum technology. Quantum processors tailored for specific applications are advancing the fields of quantum annealing, quantum simulations, quantum cryptography, quantum sensors, quantum imaging, quantum-enhanced machine learning, and more. These specialized devices have the potential to revolutionize industries, transform scientific research, and address some of the most pressing challenges in our increasingly complex world. As researchers continue to push the boundaries of quantum hardware design and quantum software development, we can anticipate even more remarkable advancements that will shape the future of quantum technology.*

Chapter 8: Quantum Cryptography and Quantum Communication Protocols

Advanced quantum cryptography protocols represent the forefront of secure communication in the age of quantum computing, where quantum technologies are harnessed to protect sensitive information from quantum threats. Quantum cryptography leverages the principles of quantum mechanics to enable secure communication channels that are immune to eavesdropping and interception. The foundation of quantum cryptography protocols lies in the unique properties of quantum states, particularly the property of quantum entanglement. Entangled particles, such as photons, are quantum-mechanically correlated in a way that their states are interdependent, even when separated by vast distances. This phenomenon allows for the creation of cryptographic keys that are distributed between two parties, Alice and Bob, in a way that any eavesdropping by a third party, Eve, can be detected. One of the pioneering quantum cryptography protocols is the BB84 protocol, introduced by Charles Bennett and Gilles Brassard in 1984. The BB84 protocol uses the properties of quantum states, specifically the polarization of photons, to create a secure key exchange between Alice and Bob. In this protocol, Alice prepares a series of photons in one of two mutually exclusive polarization states, representing binary values 0 and 1. She then sends these photons to Bob over a quantum communication channel, while Eve attempts to intercept and measure them. Bob randomly measures the polarization of each received photon using two possible measurement bases. After the transmission is complete, Alice and Bob exchange information publicly about the bases they used for encoding and measurement. They discard the measurements where the bases did not match, creating a subset of data known as the sifted key. If there is no eavesdropping, Alice and Bob's sifted keys should be identical. However, to ensure security, they perform a process called error reconciliation to correct any discrepancies caused by noise or eavesdropping. Once the error reconciliation is complete, Alice and Bob share a secure cryptographic key that can be used for encryption and decryption of their communication. The BB84 protocol provides a level of security based on the principles of quantum mechanics, as any eavesdropping attempt by Eve will inevitably disturb the quantum states of the photons, revealing her presence. Another quantum cryptography protocol that enhances security is the E91 protocol, developed by Artur Ekert in 1991. The E91 protocol relies on the phenomenon of quantum entanglement to establish a secure key between two parties. In this protocol, Alice and Bob share pairs of entangled particles, such as photons, that are in a maximally entangled state. They each measure their particles' properties, choosing randomly between two measurement bases. The measurement results are then compared to check for any discrepancies. If there is no eavesdropping, the measurement results should be correlated due to the entanglement. Alice and Bob can perform a statistical test, called the CHSH inequality test, to verify the presence of entanglement and, by extension, the absence of eavesdropping. Once they confirm the entanglement, they use the measurement outcomes to derive a secure cryptographic key. The E91 protocol offers the advantage of detecting eavesdropping without revealing any information about the key, providing a high level of security. Quantum cryptography protocols like BB84 and E91 offer quantum-resistant security, making them suitable for protecting sensitive information in a post-quantum world. These protocols ensure that even a powerful quantum computer would be unable to compromise the security of the transmitted keys. Quantum key distribution (QKD), based on these protocols, has the potential to revolutionize secure communication in various domains, including finance, government, and military applications. In addition to QKD, advanced quantum cryptography protocols explore more sophisticated methods of secure communication and cryptographic primitives. For example, quantum secure direct communication (QSDC) protocols enable the direct transmission of information without the need for shared cryptographic keys. In QSDC, Alice sends quantum states to Bob that contain the message itself, and Bob measures these states to decode the message. Any eavesdropping attempt during transmission would be immediately detectable, as it would affect the quantum states. Quantum digital signatures represent another frontier in advanced quantum cryptography. Quantum digital signatures provide a quantum-resistant alternative to classical digital signatures, ensuring the authenticity and integrity of digital messages. These signatures are generated using quantum properties, making them secure against attacks by quantum computers. Furthermore, advanced quantum cryptography protocols explore the integration of quantum cryptography with classical cryptographic techniques to enhance security further. For instance, hybrid cryptographic schemes combine quantum key distribution with classical encryption methods, providing both quantum resistance and traditional security. Hybrid schemes leverage the strengths of quantum cryptography while maintaining compatibility with existing communication infrastructure. Moreover, quantum cryptography protocols are not limited to point-to-point communication but can be extended to multi-party scenarios. Multi-party quantum cryptography protocols enable secure communication among multiple participants, ensuring that no subset of participants can collude to compromise the security of the communication. These protocols find applications in secure multi-party computations, secure auctions, and collaborative decision-making processes. As quantum technologies continue to advance, so does the development of advanced quantum cryptography protocols. The ongoing research in this field aims to enhance the security, efficiency, and practicality of quantum cryptography for real-world applications. Quantum-resistant encryption methods are essential for securing sensitive information against the potential threat of quantum computers capable of breaking classical encryption algorithms. Quantum cryptography protocols offer a promising solution to this challenge, providing a new paradigm of security that leverages the fundamental principles of quantum mechanics to protect data and communications. In an era where data privacy and security are paramount, advanced quantum cryptography protocols offer a glimpse into a future where quantum technologies safeguard our digital interactions and information. Secure quantum communication in practice represents a significant milestone in the journey towards achieving unbreakable encryption and privacy in the digital age. Quantum communication leverages the principles of quantum mechanics to enable secure communication channels that are theoretically immune to eavesdropping and interception. This emerging field of quantum technology holds the promise of transforming the way information is transmitted and protected. The foundation of secure quantum communication lies in the remarkable properties of quantum states, such as superposition and entanglement. Superposition allows quantum bits or qubits to exist in multiple states simultaneously, while entanglement establishes a profound connection between qubits, regardless of the physical separation between them. These quantum properties are the building blocks of secure quantum communication protocols and systems. Quantum key distribution (QKD) is one of the pioneering applications of secure quantum communication. QKD protocols, such as the BB84 protocol and the E91 protocol, enable two parties, Alice and Bob, to exchange cryptographic keys in a way that any eavesdropping by a third party, Eve, can be detected. In QKD, Alice generates a series of qubits, typically encoded using the polarization of photons, and sends them to Bob over a quantum channel. During transmission, Eve may attempt to intercept and measure these qubits to gain information about the key. However, the fundamental principles of quantum mechanics ensure that any eavesdropping attempt inevitably disturbs the quantum states of the qubits, revealing Eve's presence. Upon receiving the qubits, Bob performs measurements and publicly communicates the bases he used for measurement. Alice and Bob then share information about the bases they employed, allowing them to discard measurements where the bases do not match. This process results in a subset of data known as the sifted key, which should be identical for Alice and Bob if there was no eavesdropping. To ensure security, Alice and Bob perform error reconciliation to correct any discrepancies caused by noise or eavesdropping. After error reconciliation, they derive a secure cryptographic key that can be used for encryption and decryption of their communication. QKD protocols offer a level of security based on the principles of quantum mechanics, ensuring that even a powerful quantum computer would be unable to compromise the security of the transmitted keys. Moreover, QKD protocols provide security that is information-theoretically proven, meaning that it is not subject to advances in computational power or algorithmic breakthroughs. Secure quantum communication extends beyond QKD and encompasses a broader range of quantum cryptographic protocols and technologies. Quantum secure direct communication (QSDC) represents an innovative approach that enables the direct transmission of information without the need for shared cryptographic keys. In QSDC, Alice sends quantum states to Bob that contain the message itself, and Bob measures these states to decode the message. Any eavesdropping attempt during transmission would be immediately detectable, as it would affect the quantum states. Quantum digital signatures are another frontier in secure quantum communication. Quantum digital signatures provide a quantum-resistant alternative to classical digital signatures, ensuring the authenticity and integrity of digital messages. These signatures are generated using quantum properties, making them secure against attacks by quantum computers. Furthermore, secure quantum communication is not limited to point-to-point communication but can be extended to multi-party scenarios. Multi-party quantum cryptographic protocols enable secure communication among multiple participants, ensuring that no subset of participants can collude to compromise the security of the communication. These protocols find applications in secure multi-party computations, secure auctions, and collaborative decision-making processes. In addition to cryptographic protocols, secure quantum communication also encompasses the development of quantum communication networks. Quantum networks are designed to facilitate secure communication and the distribution of quantum information over large distances. These networks consist of quantum nodes that generate, manipulate, and measure quantum states, as well as quantum channels that enable the transmission of quantum information. Quantum repeaters play a critical role in quantum communication networks by extending the range of quantum communication. These devices can retransmit quantum states over long distances without compromising their security. Quantum networks are integral to the realization of a quantum internet, where quantum information can be transmitted globally with unprecedented security. Secure quantum communication in practice involves addressing various challenges and considerations. One of the practical challenges is the development of reliable quantum hardware, including sources of quantum states and quantum detectors. Quantum states must be generated with high fidelity and stability to ensure the security and integrity of quantum communication. Additionally, the transmission of quantum states over long distances can introduce losses and decoherence, which must be mitigated using quantum repeaters and error-correcting codes.

Furthermore, the integration of quantum communication into existing classical infrastructure requires careful planning and implementation. Quantum key distribution systems need to be seamlessly integrated with classical networks to enable secure communication between quantum and classical devices. Quantum communication also requires robust authentication mechanisms to ensure the trustworthiness of quantum nodes and users. Moreover, the practical deployment of secure quantum communication technologies involves regulatory and policy considerations. Governments and organizations must establish legal frameworks and standards for the use of quantum communication and encryption. These frameworks should address issues such as key management, data protection, and international cooperation in securing quantum communication networks. As secure quantum communication continues to advance, it holds the potential to revolutionize industries, protect critical infrastructure, and ensure the confidentiality of sensitive information. Quantum communication technologies are poised to play a crucial role in safeguarding the digital landscape against emerging threats, including those posed by quantum computers.

Chapter 9: Quantum Computing in Multidisciplinary Research

Quantum computing has emerged as a transformative tool in scientific research, offering the potential to address complex problems that were previously beyond the reach of classical computers. The unique properties of quantum mechanics, such as superposition and entanglement, underpin the power of quantum computing in scientific endeavors. Superposition allows quantum bits or qubits to exist in multiple states simultaneously, enabling quantum computers to explore multiple solutions to a problem in parallel. Entanglement establishes a non-classical correlation between qubits, even when separated by vast distances, enabling the creation of highly interconnected quantum systems. One of the most compelling applications of quantum computing in scientific research is in the field of quantum simulations. Quantum simulators can emulate the behavior of quantum systems with remarkable accuracy, enabling researchers to investigate complex physical and chemical phenomena. Quantum simulations have the potential to revolutionize materials science, enabling the design of novel materials with tailored properties for various applications. For example, quantum simulations can predict the behavior of exotic materials at extreme temperatures and pressures, aiding in the development of next-generation materials for energy storage, electronics, and aerospace. Quantum simulations are also instrumental in understanding quantum phase transitions, a fundamental aspect of condensed matter physics. By simulating the behavior of quantum systems at different energy levels and temperatures, researchers can gain insights into the behavior of matter *under extreme conditions. Furthermore, quantum simulations are poised to make breakthroughs in the field of quantum chemistry. Quantum computers can model the electronic structure of molecules and materials with unparalleled precision, accelerating drug discovery, and the development of new catalysts. Quantum chemistry simulations can help identify potential drug candidates and optimize their chemical properties, potentially revolutionizing pharmaceutical research. In addition to materials science and chemistry, quantum simulations have applications in fundamental physics. Researchers can use quantum simulators to investigate phenomena such as the behavior of particles in high-energy physics and the properties of exotic materials like high-temperature superconductors. Quantum simulations have the potential to unravel some of the deepest mysteries of the universe, paving the way for groundbreaking discoveries. Another area where quantum computing is making significant strides in scientific research is in optimization and combinatorial problems. Many scientific problems involve finding the best solution from a vast number of possibilities, which can be computationally intractable for classical computers. Quantum computers excel at solving optimization problems by exploring multiple solutions simultaneously. For example, quantum algorithms like the Quantum Approximate Optimization Algorithm (QAOA) have shown promise in tackling complex optimization problems in fields such as logistics, finance, and cryptography. In logistics, quantum computing can optimize supply chain management, route planning, and resource allocation, leading to cost savings and improved efficiency. In finance, quantum algorithms can be employed for portfolio optimization, risk assessment, and algorithmic trading strategies, potentially revolutionizing the financial industry. Quantum cryptography, with its focus on secure communication, also benefits from optimization algorithms, ensuring the integrity and confidentiality of quantum keys. Furthermore, quantum computing holds the potential to revolutionize machine learning and artificial intelligence. Quantum machine learning models leverage quantum processors to accelerate the training and inference processes of machine learning algorithms. These models can optimize complex functions and solve optimization problems more efficiently than classical computers. Quantum-enhanced machine learning has applications in diverse domains, including healthcare, finance, and natural language processing. In healthcare, quantum machine learning algorithms can accelerate the analysis of medical data, aiding in disease diagnosis and drug discovery. In finance, quantum machine learning can enhance predictive models for market trends and risk assessment. In natural language processing, quantum algorithms can improve language translation, sentiment analysis, and chatbot performance. Quantum computing also holds promise in the field of cryptography, where it can be used for post-quantum cryptography and the development of quantum-resistant encryption methods. With the potential threat of quantum computers breaking classical encryption algorithms, researchers are actively working on quantum-safe encryption schemes. Quantum-resistant encryption methods are designed to withstand attacks by quantum computers, ensuring the long-term security of digital communication. These encryption schemes are being developed to protect data in various domains, including secure communications, financial transactions, and data storage. Quantum computing is also contributing to advancements in cryptography by enabling the development of quantum digital signatures. Quantum digital signatures offer secure and tamper-evident authentication for digital documents* and transactions. These signatures are generated using quantum properties, making them resistant to quantum attacks. As quantum computing continues to evolve, it is becoming an indispensable tool in scientific research across diverse fields. Researchers are exploring the potential of quantum computers to tackle problems that were previously considered computationally intractable. Quantum simulations are revolutionizing materials science, chemistry, and fundamental physics by providing accurate and efficient modeling of complex systems. Optimization algorithms are enhancing logistics, finance, and cryptography, leading to more efficient and secure processes. Quantum machine learning is accelerating progress in healthcare, finance, and natural language processing, opening up new avenues for innovation. Quantum-resistant encryption methods are ensuring the security of digital communication in the post-quantum era. With ongoing advancements in quantum computing hardware and software, the impact of quantum technology on scientific research is expected to grow exponentially, offering unprecedented opportunities for discovery and innovation.

Quantum computing has transcended the boundaries of traditional disciplines and is playing a pivotal role in interdisciplinary projects that span a wide spectrum of scientific and technological domains. The unique capabilities of quantum computers, rooted in the principles of quantum mechanics, make them a powerful tool for addressing complex problems that were once deemed insurmountable. One of the remarkable features of quantum computing is its versatility, as it can be applied to an array of interdisciplinary projects, from materials science and drug discovery to climate modeling and artificial intelligence. In materials science, quantum computing is revolutionizing the way new materials are discovered and optimized. Quantum simulators can accurately model the behavior of atoms and molecules, allowing researchers to predict material properties with unparalleled precision. This capability is invaluable in designing materials with tailored properties for various applications, including renewable energy, electronics, and aerospace. Interdisciplinary projects in materials science benefit from quantum computing's ability to explore the quantum behavior of materials at different temperatures, pressures, and energy levels. By simulating the behavior of materials under extreme conditions, researchers can unlock novel insights into their properties and performance. Furthermore, quantum computers are making significant contributions to the field of drug discovery and development. The simulation of molecular interactions and chemical reactions is a computationally intensive task that quantum computers are well-suited to tackle. Quantum chemistry simulations can accelerate the identification of potential drug candidates, predict their interactions with biological targets, and optimize their chemical structures. This has the potential to expedite the drug discovery process, leading to the development of new treatments for diseases and medical conditions. Quantum computing's role in interdisciplinary projects extends to climate modeling and environmental research. Climate models require vast computational resources to simulate the complex interactions between various factors affecting the Earth's climate. Quantum computers offer the potential to enhance the accuracy and efficiency of climate models by simulating the behavior of atmospheric molecules, ocean currents, and greenhouse gases with greater precision. These simulations can provide insights into climate change, extreme weather events, and the impact of human activities on the environment. Interdisciplinary projects in climate science can benefit from quantum computing's ability to analyze large datasets and *perform complex simulations, enabling more accurate predictions and informed policy decisions. In the realm of artificial intelligence (AI) and machine learning, quantum computing is poised to revolutionize the field. Quantum machine learning models leverage quantum processors to accelerate the training and inference processes of AI algorithms. These models can optimize complex functions, solve optimization problems, and process large datasets more efficiently than classical computers. Interdisciplinary projects in AI and machine learning can harness quantum computing's power to enhance natural language processing, image recognition, and predictive analytics. This has implications for fields such as healthcare, finance, and autonomous systems, where AI plays a pivotal role in decision-making and data analysis. Moreover, quantum computing is contributing to interdisciplinary projects focused on cryptography and data security. The advent of powerful quantum computers poses a potential threat to classical encryption algorithms. As a result, researchers are developing quantum-resistant encryption methods to protect sensitive data in various domains. Interdisciplinary projects in cybersecurity can leverage quantum computing to design and implement quantum-safe encryption schemes that are resilient to attacks by quantum computers. These encryption methods offer long-term security and data protection, ensuring the confidentiality and integrity of digital communication. Quantum digital signatures, generated using quantum properties, provide a tamper-evident mechanism for authenticating digital documents and transactions in interdisciplinary projects related to secure communications and data integrity. Furthermore, quantum computing's role in interdisciplinary projects extends to optimization and combinatorial problems. Many real-world problems involve finding the best solution from a vast* number of possibilities, a task that can be computationally intractable for classical computers. Quantum algorithms, such as the Quantum Approximate Optimization Algorithm (QAOA), excel at solving optimization problems by exploring multiple solutions simultaneously. Interdisciplinary projects in logistics, finance, and cryptography can benefit from quantum computing's ability to optimize supply chain management, portfolio allocation, and cryptographic key distribution. Quantum-enhanced optimization algorithms offer the potential to revolutionize decision-making processes in these domains. Quantum computing's interdisciplinary impact is further amplified by its potential to address challenges in fields such as energy, transportation, and materials engineering. For example, quantum computing can optimize the design and operation of renewable energy systems, leading to more efficient solar panels, wind turbines, and energy storage solutions. In the transportation sector, quantum computing can enhance route planning, traffic management, and resource allocation for autonomous vehicles and smart transportation networks. Interdisciplinary projects in materials engineering can leverage quantum computing to design advanced materials with tailored properties for structural engineering, aerospace, and nanotechnology. Moreover, quantum computing is fostering collaboration between researchers from different disciplines, creating a synergy that accelerates innovation and problem-solving. Interdisciplinary teams are harnessing quantum computing's capabilities to address complex challenges that require expertise from multiple domains. These collaborations are breaking down traditional silos and fostering a culture of interdisciplinary research that transcends the boundaries of individual disciplines. Institutions and organizations are recognizing the importance of interdisciplinary projects that incorporate quantum computing as a central tool for tackling complex, real-world problems. Funding agencies and research institutions are promoting cross-disciplinary research initiatives that harness the potential of quantum computing to advance scientific understanding and drive technological innovation. Quantum computing's role in interdisciplinary projects is not limited to academia and research institutions but extends to industry and practical applications. Companies and organizations are exploring how quantum computing can revolutionize their operations and processes, from optimizing supply chains and logistics to accelerating drug discovery and materials design. The quantum ecosystem is expanding to support interdisciplinary projects, with quantum cloud computing platforms and quantum software development tools becoming increasingly accessible to researchers and practitioners across various fields. In summary, quantum computing's role in interdisciplinary projects is transformative, offering new avenues for innovation and problem-solving across a wide range of scientific and technological domains. The unique capabilities of quantum computers, including quantum simulations, optimization algorithms, and quantum-enhanced machine learning, are revolutionizing materials science, drug discovery, climate modeling, artificial intelligence, cybersecurity, and more. Interdisciplinary collaboration and interdisciplinary projects are at the forefront of harnessing the power of quantum computing to address complex, real-world challenges and drive advancements in science and technology.

Chapter 10: Quantum Computing Beyond Moore's Law

Pushing the limits of quantum computing is at the forefront of scientific and technological research, as scientists and engineers strive to unlock the full potential of quantum processors. Quantum computing leverages the principles of quantum mechanics, which enable quantum bits or qubits to exist in multiple states simultaneously, offering an exponential increase in computational power compared to classical bits. This unique feature of quantum computing has the potential to revolutionize fields such as materials science, cryptography, optimization, and artificial intelligence. One of the primary challenges in pushing the limits of quantum computing is achieving and maintaining quantum coherence, the property that allows qubits to maintain their superposition and entanglement. Quantum coherence is fragile and can be easily disrupted by environmental factors such as temperature, electromagnetic radiation, and even cosmic rays. Researchers are continuously developing techniques and technologies to extend the coherence times of qubits, enabling longer and more complex quantum computations. Another critical aspect of advancing quantum computing is increasing the number of qubits and improving their quality. Quantum processors with a greater number of qubits can tackle more complex problems and perform more intricate calculations. Furthermore, qubits must be highly reliable, with low error rates, to ensure the accuracy and precision of quantum computations. Researchers are exploring various qubit technologies, including superconducting qubits, trapped ions, and topological qubits, each with its own advantages and challenges. In the quest to push the limits of quantum computing, error correction plays a pivotal role. Quantum error correction codes, such as the surface code, are essential for mitigating errors that occur

221

during quantum computations. These codes enable the detection and correction of errors without disrupting the quantum states of qubits. As quantum processors become more powerful, the need for efficient and scalable error correction becomes increasingly critical. Developing fault-tolerant quantum computing, where errors are suppressed to negligible levels, is a significant milestone in pushing the limits of quantum computing. Quantum algorithms are a driving force in expanding the horizons of quantum computing. Researchers are actively designing and optimizing quantum algorithms that can solve real-world problems faster and more efficiently than classical algorithms. For example, quantum algorithms like Grover's search algorithm and Shor's factoring algorithm have the potential to impact cryptography and data encryption significantly. By breaking classical encryption methods, quantum computers pose a threat to data security, making the development of quantum-resistant encryption schemes a top priority. Pushing the limits of quantum computing also involves exploring the potential of quantum machine learning. Quantum machine learning models leverage quantum processors to accelerate the training and inference processes of machine learning algorithms. These models can process and analyze vast datasets more efficiently, leading to advancements in fields such as healthcare, finance, and natural language processing. Quantum-enhanced machine learning has the potential to revolutionize predictive analytics and decision-making processes. Quantum computing's role in pushing the limits of scientific discovery is particularly evident in the field of quantum simulations. Quantum simulators can emulate the behavior of quantum systems with remarkable accuracy, allowing researchers to investigate complex physical and chemical phenomena. In materials science, quantum simulations can predict the properties of novel materials, paving the way for the development of advanced materials with tailored characteristics. Quantum simulations are also instrumental in quantum chemistry, where they can model the electronic structure of molecules and materials, accelerating drug discovery and materials design. Moreover, quantum simulations have applications in fundamental physics, enabling researchers to explore the behavior of particles in high-energy physics and the properties of exotic materials like high-temperature superconductors. Pushing the limits of quantum computing is closely tied to the development of quantum hardware. Quantum processors must be continually improved to accommodate larger qubit counts, reduce error rates, and extend quantum coherence times. This requires advancements in materials science, fabrication techniques, and cryogenic engineering. Superconducting qubits, for instance, are typically operated at ultra-low temperatures to maintain their quantum coherence. Developing reliable and scalable quantum hardware is a fundamental aspect of pushing the boundaries of quantum computing. Quantum networking is another frontier in quantum computing that extends its capabilities beyond individual quantum processors. Quantum networks enable the distribution of entangled qubits over long distances, facilitating secure communication and quantum key distribution. These networks are integral to the realization of a quantum internet, where quantum information can be transmitted globally with unprecedented security. Quantum repeaters play a crucial role in extending the range of quantum communication by mitigating losses and maintaining quantum entanglement. Pushing the limits of quantum computing involves building and expanding quantum networks to enable global-scale quantum communication. Furthermore, the interdisciplinary nature of quantum computing is driving collaboration between researchers from diverse fields. Interdisciplinary projects harness the power of quantum computing to address complex, real-world challenges that require expertise from multiple domains. These collaborations are breaking down traditional silos and fostering a culture of interdisciplinary research that *transcends the boundaries of individual disciplines. Pushing the limits of quantum computing is not limited to academia and research institutions but extends to industry and practical applications. Companies and organizations are exploring how quantum computing can revolutionize their operations and processes, from optimizing supply chains and logistics to accelerating drug discovery and materials design. Quantum computing has the potential to disrupt existing industries and create new opportunities for innovation. The quantum ecosystem is expanding to support the growing demand for quantum computing resources and expertise. Quantum cloud computing platforms and quantum software development tools are becoming increasingly accessible to researchers and practitioners across various fields. In summary, pushing the limits of quantum computing is a multifaceted endeavor that encompasses advancing quantum hardware, optimizing quantum algorithms, and expanding the scope of interdisciplinary projects. Researchers, engineers, and organizations are collaboratively driving the field of quantum computing forward to unlock its full potential and address some of the most pressing challenges in science and technology. Quantum technologies represent a paradigm shift in the world of science and engineering, offering capabilities that transcend the constraints of classical technologies. Classical physics, which governs the behavior of everyday objects, has served as the foundation for numerous technological advancements over centuries. However, as we venture into the realm of quantum physics, we encounter a new set of rules and phenomena that open up unprecedented possibilities. The heart of quantum technologies lies in the principles of quantum mechanics, a branch of physics that deals with the behavior of particles at the quantum level. At this scale, the classical distinction between particles and waves blurs, and particles can exist in multiple states simultaneously, a phenomenon known as superposition. This inherent superposition property allows* quantum bits, or qubits, to represent and process information in a radically different way from classical bits. In classical computing, information is stored as binary code, where each bit can be either a 0 or a 1. Quantum computing, on the other hand, harnesses the power of qubits, which can exist in a superposition of 0 and 1 states, enabling quantum computers to perform complex calculations exponentially faster than classical computers. This fundamental difference is at the core of quantum technologies and has far-reaching implications. Quantum technologies encompass a wide range of applications, from quantum computing and quantum cryptography to quantum sensing and quantum communication. One of the most anticipated quantum technologies is quantum computing, which has the potential to revolutionize fields such as materials science, drug discovery, and optimization. Quantum algorithms, designed to leverage the unique properties of qubits, promise to solve complex problems that were previously considered computationally intractable. For example, quantum computers can simulate the behavior of molecules and materials at the quantum level, enabling the discovery of novel materials with tailored properties for various applications. They can also accelerate the process of drug discovery by modeling the interactions of molecules and predicting their effectiveness as pharmaceuticals. Moreover, quantum computers excel at optimization problems, offering solutions that can revolutionize supply chain management, financial modeling, and cryptographic key distribution. Quantum cryptography, another quantum technology, addresses the need for secure communication in the digital age. Classical cryptographic methods, while effective, could be vulnerable to attacks by future quantum computers. Quantum cryptography leverages the principles of quantum mechanics to provide provably secure communication channels. Quantum key distribution (QKD) protocols, such as the BB84 protocol and the E91 protocol, enable two parties to exchange cryptographic keys in a way *that any eavesdropping attempt is immediately detectable. This ensures the confidentiality and integrity of data transmission. Quantum secure direct communication (QSDC) takes quantum cryptography a step further, allowing for direct communication of messages without the need for shared cryptographic keys. This is achieved by encoding the message itself into quantum states, making it inherently secure against eavesdropping. Quantum technologies also play a crucial role in quantum sensing, enabling highly precise measurements and detection capabilities beyond the reach of classical sensors. Quantum sensors can detect minute changes in physical properties such as temperature, magnetic fields, and gravitational forces. This has applications in a wide range of fields, including geology, navigation, and medical imaging. For example, quantum magnetometers can detect underground mineral deposits, while quantum gravimeters can map variations in gravitational fields for geological exploration. Quantum-enhanced imaging techniques offer higher resolution and sensitivity for medical diagnostics and non-destructive testing. Furthermore, quantum technologies are poised to revolutionize communication networks with the development of quantum communication. Quantum networks enable secure transmission of information using entangled qubits, even over long distances. These networks are critical for the realization of a quantum internet, where quantum information can be transmitted globally with unmatched security. Quantum repeaters are key components of quantum networks, allowing the distribution of entangled qubits over extended distances without compromising their quantum properties. This opens up possibilities for secure quantum communication and the development of quantum-enhanced protocols for tasks such as secure multi-party computations and quantum-enhanced distributed computing. Quantum technologies also have a significant impact on the field of quantum simulation, enabling researchers to model and study complex quantum systems with unprecedented accuracy.*

Quantum simulators can emulate the behavior of quantum systems, ranging from the behavior of atoms and molecules to the properties of exotic materials and high-energy physics phenomena. This has applications in materials science, quantum chemistry, and condensed matter physics. For instance, quantum simulations can predict the behavior of materials under extreme conditions, aiding in the development of next-generation materials for energy storage, electronics, and aerospace. In addition to these applications, quantum technologies are advancing interdisciplinary research projects that leverage the unique capabilities of quantum systems. Researchers from diverse fields are collaborating to explore how quantum technologies can address complex problems that require expertise from multiple domains. Interdisciplinary projects in materials science, drug discovery, climate modeling, and artificial intelligence are benefiting from quantum computing's computational power and quantum simulations' accuracy. These collaborations are breaking down traditional silos and fostering a culture of interdisciplinary research that transcends the boundaries of individual disciplines. As quantum technologies continue to advance, they hold the promise of solving some of the most pressing challenges in science, technology, and society.

BOOK 4
QUANTUM COMPUTING
A MULTIDISCIPLINARY APPROACH FOR EXPERTS
ROB BOTWRIGHT

Chapter 1: Quantum Mechanics and Advanced Quantum Theory

Quantum mechanics, often referred to as quantum physics or quantum theory, is a fundamental branch of physics that provides a framework for understanding the behavior of matter and energy at the smallest scales. It is a cornerstone of modern physics, revolutionizing our understanding of the physical world and challenging classical physics, which primarily governs the behavior of macroscopic objects. Quantum mechanics introduces a new set of principles and phenomena that are distinctly different from classical physics, and it plays a crucial role in explaining the behavior of atoms, molecules, subatomic particles, and the universe at large. One of the central tenets of quantum mechanics is the wave-particle duality, which asserts that particles like electrons and photons can exhibit both particle-like and wave-like characteristics depending on the experimental context. This duality challenges our intuitive understanding of physical objects, as particles can exhibit interference patterns similar to waves in certain experiments. In addition to wave-particle duality, another foundational concept of quantum mechanics is the uncertainty principle, famously formulated by Werner Heisenberg. The uncertainty principle states that there are intrinsic limits to how precisely we can simultaneously know certain pairs of properties of a particle, such as its position and momentum. This principle introduces a fundamental level of unpredictability into quantum systems, setting them apart from classical deterministic systems. Quantum mechanics also introduces the concept of quantization, where physical properties, such as energy levels in atoms and molecules, are restricted to discrete, quantized values rather than continuous ones. For instance, the energy levels of electrons in an atom are quantized, and electrons can only occupy specific energy states, known as quantum states. This quantization leads to phenomena like the emission and absorption of discrete energy quanta, known as photons. Furthermore, quantum mechanics introduces the notion of superposition, a fundamental principle that allows quantum systems to exist in a linear combination of multiple states simultaneously. Superposition leads to the intriguing phenomenon of entanglement, where the quantum states of two or more particles become correlated in such a way that the properties of one particle are instantaneously linked to the properties of the other(s), even when separated by vast distances. Albert Einstein famously referred to this phenomenon as "spooky action at a distance." Quantum mechanics also features the concept of probability amplitudes, which are complex numbers associated with quantum states. These amplitudes provide a way to calculate the probability of observing a particular outcome when measuring a quantum system. The Born Rule, formulated by Max Born, relates the square of the absolute value of the probability amplitudes to the probability of observing a specific measurement outcome. This probabilistic nature of quantum mechanics is fundamentally different from classical physics, where determinism reigns. The mathematical framework of quantum mechanics relies heavily on linear algebra and the use of wave functions to describe quantum states. A wave function is a mathematical function that encapsulates all the information about a quantum system, including its position, momentum, and other properties. The Schrödinger equation, developed by Erwin Schrödinger, serves as the fundamental equation of quantum mechanics, describing how the wave function of a quantum system evolves over time. Solving the Schrödinger equation for a particular quantum system allows us to determine the quantum states and predict their future behavior. Quantum mechanics also encompasses the concept of operators, which are mathematical constructs used to represent physical observables in quantum systems, such as position, momentum, and angular momentum. Operators play a central role in quantum mechanics, as they act on wave functions to yield observable quantities. Heisenberg's matrix mechanics and Schrödinger's wave mechanics, developed independently in the 1920s, are two equivalent mathematical formalisms that describe quantum mechanics. Both formalisms have been crucial in the development of quantum theory and its applications. Quantum mechanics has been rigorously tested through countless experiments, and its predictions have been verified with remarkable accuracy. From the behavior of electrons in atoms to the properties of subatomic particles like quarks and neutrinos, quantum mechanics has consistently provided a robust framework for understanding the physical world at the quantum level. The Copenhagen interpretation, formulated by Niels Bohr and Werner Heisenberg, is one of the most widely known interpretations of quantum mechanics. It posits that quantum systems do not have definite properties until they are measured, and the act of measurement collapses the quantum state into one of the possible measurement outcomes. This interpretation highlights the role of the observer in quantum phenomena and has been a subject of philosophical debate for decades. Another interpretation, the Many-Worlds interpretation proposed by Hugh Everett III, suggests that all possible measurement outcomes occur in separate branches of a constantly branching universe. In this view, the quantum superposition of states is not collapsed but leads to the existence of multiple parallel universes. The Many-Worlds interpretation offers a different perspective on quantum measurement and has gained its share of proponents and critics. Quantum mechanics has a wide range of applications, extending from the microscopic world of atoms and particles to macroscopic systems like superconductors and superfluids. Quantum mechanics is the foundation of quantum chemistry, which plays a crucial role in understanding chemical reactions and the behavior of molecules. As we delve deeper into the realm of quantum theory, we encounter advanced topics that push the boundaries of our understanding of the quantum world. These topics build upon the foundational principles of quantum mechanics and introduce intriguing concepts that challenge our classical intuitions. One such advanced topic is the concept of quantum entanglement, which Einstein famously referred to as "spooky action at a distance." Entanglement is a phenomenon in which two or more quantum particles become correlated in such a way that the properties of one particle are instantaneously linked to the properties of the other(s), even when separated by vast distances. This phenomenon defies classical notions of locality and has been experimentally confirmed through tests of Bell's inequalities. Quantum entanglement plays a crucial role in various quantum technologies, such as quantum cryptography and quantum teleportation. Another advanced topic in quantum theory is quantum measurement and the role of the observer. The act of measuring a quantum system can lead to the collapse of its quantum state into one of the possible measurement outcomes. This raises profound questions about the nature of reality and the role of consciousness in the quantum world. The Copenhagen interpretation, which suggests that quantum systems do not have definite properties until measured, highlights the observer's influence. However, alternative interpretations, like the

Many-Worlds interpretation, propose that all possible measurement outcomes occur in separate branches of a constantly branching universe. This debate continues to captivate physicists and philosophers alike. Quantum decoherence is another advanced topic that explores the interaction between quantum systems and their environments. Decoherence arises from the entanglement of a quantum system with its surrounding environment, causing the quantum coherence to degrade over time. This phenomenon plays a significant role in the transition from the quantum realm to the classical world and presents challenges for quantum computing and quantum communication. Researchers are actively working on strategies to mitigate decoherence and preserve quantum states. Advanced quantum mechanics also delves into the concept of quantum superposition, where quantum systems can exist in a linear combination of multiple states simultaneously. This property is fundamental to quantum computers, allowing them to perform complex calculations exponentially faster than classical computers. Quantum algorithms, like Shor's algorithm and Grover's search algorithm, leverage superposition to solve problems that would be infeasible for classical computers. Furthermore, quantum superposition enables the development of quantum-enhanced sensors and detectors, offering unprecedented precision in measurements. Quantum theory also introduces the concept of quantum states and their mathematical representation using wave functions. Wave functions describe the probability amplitudes associated with quantum states, and they obey the Schrödinger equation, which governs the evolution of quantum systems over time. Understanding and manipulating quantum states are essential for quantum computation and quantum simulation. Advanced topics in quantum theory include the study of quantum field theory and its applications in high-energy physics and particle physics. Quantum field theory combines the principles of quantum mechanics with special relativity to describe the behavior of quantum fields and particles. It has been instrumental in our understanding of the fundamental forces of nature, such as electromagnetism and the strong and weak nuclear forces. Quantum field theory also gave rise to the prediction of particle interactions, like the Higgs boson's discovery. Furthermore, quantum field theory plays a vital role in cosmology, describing the behavior of the early universe during the Big Bang. Another advanced topic is quantum information theory, which explores the principles of quantum computation, quantum communication, and quantum cryptography. Quantum information theory deals with the transmission and manipulation of quantum information using quantum bits or qubits. Quantum algorithms, developed within this framework, aim to solve problems efficiently by harnessing quantum parallelism and entanglement. Quantum communication protocols, such as quantum key distribution (QKD), provide secure methods for transmitting information over long distances. Quantum cryptography utilizes the fundamental principles of quantum mechanics to ensure the confidentiality and integrity of data transmission. Advanced topics in quantum theory also encompass the study of quantum optics, where the behavior of photons and their interactions with matter are explored. Quantum optics has led to groundbreaking experiments, such as the double-slit experiment, which highlights the wave-particle duality of photons. It has also paved the way for the development of technologies like quantum lasers and quantum sensors. The concept of quantum teleportation is another intriguing advanced topic. Quantum teleportation allows the transfer of quantum information from one location to another without physical transmission. It relies on entanglement and the measurement of quantum states to reconstruct the exact quantum state at a distant location. While quantum teleportation has been successfully demonstrated in experiments, its practical applications remain a subject of ongoing research. Advanced quantum theory also explores the concept of quantum computing models and architectures. These models include adiabatic quantum computing, quantum annealing, and topological quantum computing, each with its unique approach to solving complex problems. Quantum hardware platforms, such as superconducting qubits and trapped ions, are developed to support these models. Quantum error correction codes and fault-tolerant quantum computing are essential for building reliable and scalable quantum computers. In summary, advanced topics in quantum theory expand upon the foundational principles of quantum mechanics, introducing concepts like entanglement, quantum measurement, and decoherence. These topics play a pivotal role in quantum technologies, quantum information theory, quantum optics, and quantum computing. They challenge our classical intuitions and continue to drive innovation and discovery in the field of quantum physics.

***Chapter 2: Quantum Information Science: Foundations and
Extensions***

*In the realm of quantum information, several fundamental
concepts serve as the building blocks of this rapidly evolving
field. These concepts underpin the development of quantum
technologies, including quantum computing, quantum
cryptography, and quantum communication. Quantum bits,
or qubits, are at the heart of quantum information. Unlike
classical bits, which can be either 0 or 1, qubits can exist in a
superposition of both 0 and 1 states simultaneously. This
unique property enables quantum computers to perform
parallel computations and tackle complex problems with
exponential speedup. Qubits can be realized using various
physical systems, including superconducting circuits, trapped
ions, and photons. Entanglement, another foundational
concept, describes the strong correlation between quantum
particles, such that the properties of one particle are
instantaneously linked to those of another, even when
separated by vast distances. Entanglement plays a pivotal
role in quantum cryptography and quantum teleportation.
Quantum cryptography leverages the principles of quantum
mechanics to provide secure communication channels. One
of its core protocols is quantum key distribution (QKD),
where two parties can securely exchange cryptographic keys
without the risk of eavesdropping. Quantum teleportation,
on the other hand, allows the transfer of quantum
information from one location to another without physical
transmission. This process relies on entanglement and the
measurement of quantum states. The no-cloning theorem is
a fundamental principle in quantum information theory,
stating that it is impossible to create an identical copy of an*

arbitrary unknown quantum state. This theorem has implications for secure communication, as it prevents an eavesdropper from making an undetectable copy of a transmitted quantum key. Quantum channels represent the medium through which quantum information is transmitted. They encompass various physical systems, including optical fibers and free-space communication. Quantum communication protocols aim to protect quantum information from noise and decoherence during transmission. Quantum superposition allows quantum systems to exist in a linear combination of multiple states. This property is harnessed in quantum algorithms, which exploit superposition to perform computations more efficiently than classical algorithms. Shor's algorithm, for instance, can factor large numbers exponentially faster than classical methods, posing a threat to classical encryption schemes. Grover's search algorithm accelerates database searches and has applications in optimization. Quantum gates are the quantum analogs of classical logic gates, performing operations on qubits. Common quantum gates include the Hadamard gate, the Pauli-X gate, and the CNOT gate. Quantum circuits combine these gates to implement quantum algorithms. The development of quantum error correction codes is crucial for building fault-tolerant quantum computers. These codes detect and correct errors that can occur during quantum computations. The surface code is one such error-correcting code, offering promising solutions to error mitigation in quantum processors. Quantum supremacy refers to the point at which a quantum computer can perform a specific task faster than the most advanced classical supercomputer. Achieving quantum supremacy is a significant milestone in the field of quantum computing. Quantum annealing is a quantum computing approach that focuses on optimization problems. It leverages *quantum fluctuations to explore energy landscapes and find optimal solutions. D-Wave Systems is a prominent company known for its quantum annealers. Quantum algorithms, such as the Quantum Approximate Optimization Algorithm (QAOA), are designed to tackle optimization problems efficiently. Quantum machine learning combines quantum computing with machine learning techniques to process and analyze large datasets more efficiently. Quantum-enhanced algorithms have the potential to revolutionize fields like drug discovery and financial modeling. Quantum simulation allows researchers to emulate and study quantum systems with unprecedented accuracy. This has applications in materials science, quantum chemistry, and high-energy physics. Quantum-enhanced sensors offer increased precision in measurements, with applications in fields like navigation and medical imaging. Quantum-enhanced imaging techniques can provide higher resolution and sensitivity. Quantum-enhanced spectroscopy can detect minute variations in physical properties, such as temperature and magnetic fields. Quantum-enhanced clocks provide highly accurate timekeeping, with potential applications in global positioning systems (GPS) and scientific experiments. Quantum-enhanced metrology allows for more precise measurements of physical quantities, benefiting fields like geology and navigation. Quantum-enhanced imaging techniques offer advantages in medical diagnostics and non-destructive testing. Quantum-enhanced microscopy enables the imaging of nanoscale structures and biological molecules with higher resolution. Quantum-enhanced communication networks leverage the principles of quantum mechanics to enable secure communication over long distances. Quantum key distribution (QKD) protocols provide a secure means of transmitting cryptographic keys. Quantum repeaters extend the range of quantum communication by mitigating losses* and preserving entanglement. Quantum-enhanced distributed computing harnesses the power of entanglement for tasks like secure multi-party computations. Quantum-enhanced cryptography is essential for ensuring the confidentiality and integrity of data transmission in quantum communication. Quantum-enhanced algorithms and protocols have the potential to revolutionize fields like cybersecurity and data privacy. Quantum-enhanced sensors and detectors offer unparalleled precision in various fields, including geology and medical imaging. Quantum-enhanced navigation systems can provide more accurate positioning and timing information. Quantum-enhanced imaging techniques have applications in medical diagnostics and materials characterization.

Expanding the frontiers of quantum information science is an exciting endeavor that promises to reshape our technological landscape. This cutting-edge field explores the fundamental principles of quantum mechanics and harnesses them to develop revolutionary technologies and applications. At its core, quantum information science seeks to exploit the unique properties of quantum systems to perform tasks that are impossible or prohibitively difficult with classical systems. One of the foundational concepts in quantum information science is the qubit, a quantum counterpart to classical bits. Qubits, unlike classical bits that are either in a state of 0 or 1, can exist in a superposition of both states simultaneously. This property enables quantum computers to process information exponentially faster than classical computers, making them ideal for solving complex problems in fields such as cryptography, optimization, and materials science. The second cornerstone of quantum information science is quantum entanglement, a phenomenon where the properties of two or more quantum particles become intrinsically linked, regardless of the physical distance separating them.

Entanglement is the basis for secure quantum communication, as it allows for the creation of cryptographic keys that are inherently secure against eavesdropping. This capability is a game-changer in the world of cybersecurity, where the protection of sensitive information is paramount. Moreover, entanglement enables quantum teleportation, a process that allows quantum information to be transmitted from one location to another without physically traversing the space in between. Quantum teleportation may sound like science fiction, but it is a real and demonstrated phenomenon with profound implications for long-distance communication and quantum networking. The no-cloning theorem, a fundamental principle in quantum information science, states that it is impossible to create an exact copy of an arbitrary unknown quantum state. This theorem has significant implications for quantum cryptography, as it ensures that a quantum key cannot be copied or intercepted by an eavesdropper without detection. Quantum channels are the conduits through which quantum information is transmitted, encompassing various physical systems like optical fibers and free-space communication. Quantum communication protocols are designed to protect quantum information from noise and decoherence during transmission, ensuring the reliable exchange of quantum keys and information. Quantum superposition, another pivotal concept, allows quantum systems to exist in a linear combination of multiple states simultaneously. This property forms the basis of quantum algorithms, which can tackle problems with unprecedented efficiency. Shor's algorithm, for instance, can factor large numbers exponentially faster than classical algorithms, posing a threat to classical encryption schemes. Grover's search algorithm accelerates database searches and has applications in optimization. Quantum gates are the quantum counterparts of classical logic gates, performing operations on qubits. These gates are the building blocks of quantum circuits, where they are combined to implement quantum algorithms. The Hadamard gate, the Pauli-X gate, and the CNOT gate are some common examples of quantum gates. Quantum circuits are the circuit diagrams that represent the flow of quantum information through a series of quantum gates, illustrating the operations performed on qubits. These circuits are crucial for understanding and designing quantum algorithms. Quantum error correction codes are essential for building reliable and scalable quantum computers. These codes detect and correct errors that can occur during quantum computations, preserving the integrity of quantum information. The development of quantum error correction is a critical step towards achieving fault-tolerant quantum computation. Quantum supremacy, a significant milestone in quantum information science, refers to the point at which a quantum computer can outperform the most advanced classical supercomputer in a specific task. Reaching quantum supremacy demonstrates the potential of quantum computing to revolutionize fields like materials science, cryptography, and optimization. Quantum annealing is an approach to quantum computing that focuses on optimization problems. It leverages quantum fluctuations to explore energy landscapes and find optimal solutions. Companies like D-Wave Systems are known for their quantum annealers, which have applications in fields such as finance and logistics. Quantum algorithms, such as the Quantum Approximate Optimization Algorithm (QAOA), are designed to tackle optimization problems efficiently, making them valuable tools for industry and research. Quantum machine learning is an interdisciplinary field that combines quantum computing with machine learning techniques to process and analyze large datasets more efficiently.

Quantum-enhanced algorithms have the potential to revolutionize fields like drug discovery, financial modeling, and artificial intelligence. Quantum simulation allows researchers to emulate and study complex quantum systems with unprecedented accuracy. This has applications in materials science, quantum chemistry, and condensed matter physics. Quantum-enhanced sensors and detectors offer unparalleled precision in measurements, with applications in fields like geology, navigation, and medical imaging. Quantum-enhanced imaging techniques provide higher resolution and sensitivity, enabling advancements in medical diagnostics and non-destructive testing. Quantum-enhanced spectroscopy can detect minute variations in physical properties, such as temperature and magnetic fields, making it invaluable for scientific research and industry applications. Quantum-enhanced clocks provide highly accurate timekeeping, with potential applications in global positioning systems (GPS) and scientific experiments. Quantum-enhanced metrology allows for more precise measurements of physical quantities, benefiting fields like geology, navigation, and manufacturing. Quantum-enhanced imaging techniques offer advantages in medical diagnostics and materials characterization, paving the way for more effective healthcare and materials research. Quantum-enhanced navigation systems can provide more accurate positioning and timing information, with applications in autonomous vehicles and logistics. Quantum-enhanced microscopy enables the imaging of nanoscale structures and biological molecules with higher resolution, advancing fields like biology and materials science.

Chapter 3: Quantum Algorithms: Beyond the Quantum Speedup

In the ever-evolving landscape of quantum computing, advanced quantum algorithm design stands as a beacon of innovation, pushing the boundaries of what is computationally achievable in ways that classical computers can only dream of. These quantum algorithms are the driving force behind the transformative power of quantum computers, promising exponential speedup for solving complex problems. At the heart of advanced quantum algorithm design lies the exploitation of quantum parallelism, a feature inherent to quantum systems. Unlike classical computers that process information sequentially, quantum computers leverage quantum bits, or qubits, to explore multiple possibilities simultaneously. This inherent parallelism enables quantum algorithms to outperform classical counterparts dramatically in specific applications. One of the most celebrated quantum algorithms is Shor's algorithm, a groundbreaking achievement in the realm of number theory and cryptography. Shor's algorithm has the remarkable capability to factor large numbers exponentially faster than the best-known classical algorithms. This poses a significant threat to classical encryption methods that rely on the difficulty of factoring large numbers, such as RSA encryption. The implications of Shor's algorithm have spurred interest in post-quantum cryptography, seeking cryptographic solutions that remain secure even in the age of quantum computers. Quantum simulation algorithms are another facet of advanced quantum algorithm design. They enable the emulation and exploration of complex quantum systems, a task that classical computers struggle to perform efficiently. Quantum simulators hold immense promise for fields like quantum chemistry, condensed matter physics, and materials science. By accurately simulating quantum interactions at the quantum level, researchers can gain insights into the behavior of molecules, materials, and exotic quantum phases that were previously computationally intractable. One such algorithm, the Variational Quantum Eigensolver (VQE), allows for the efficient calculation of molecular properties, revolutionizing the way chemists design new drugs and materials. Grover's search algorithm is yet another remarkable quantum algorithm that showcases the prowess of advanced quantum algorithm design. Grover's algorithm exponentially accelerates the process of searching unsorted databases or solving unstructured search problems. In a classical computer, searching through an unsorted database typically requires checking each entry individually, which takes linear time. Grover's algorithm, on the other hand, can perform this task with only a quadratic number of queries, offering a quadratic speedup that becomes increasingly advantageous for large datasets. This has applications in various domains, including database management and optimization. Quantum machine learning algorithms represent an exciting intersection of quantum computing and artificial intelligence. These algorithms aim to harness the computational power of quantum computers to process and analyze vast datasets more efficiently than classical counterparts. Quantum machine learning promises to revolutionize industries such as finance, healthcare, and natural language processing. The Quantum Support Vector Machine (QSVM) is one example of a quantum machine learning algorithm that seeks to enhance classification tasks by exploiting quantum parallelism. Quantum-enhanced optimization algorithms are another area of focus within advanced quantum algorithm design. These algorithms aim to solve complex optimization problems that are ubiquitous in fields like logistics, finance, and engineering. The Quantum Approximate Optimization Algorithm (QAOA) is a notable example that combines classical and quantum optimization techniques to find approximate solutions to challenging problems. QAOA leverages the quantum parallelism of qubits to explore a broad solution space efficiently. Quantum algorithms for linear algebra are indispensable tools for various scientific and engineering applications. These algorithms streamline the process of solving linear equations, matrix inversion, and eigenvalue problems, which are fundamental in physics, engineering, and data analysis. The Quantum Phase Estimation (QPE) algorithm, for instance, provides an exponential speedup for finding eigenvalues of unitary operators, making it invaluable for quantum chemistry simulations and quantum cryptography protocols. Quantum algorithms for graph theory and network analysis have emerged as powerful tools for solving complex problems related to networks, such as finding optimal routes or detecting anomalies. These algorithms leverage quantum principles to explore graph structures efficiently, offering promising applications in transportation, logistics, and cybersecurity. Quantum algorithms for combinatorial optimization are well-suited for solving problems involving the selection of the best combination from a finite set of options. They have applications in resource allocation, scheduling, and portfolio optimization, with the Quantum Approximate Optimization Algorithm (QAOA) leading the way in solving such problems efficiently. Quantum algorithms for machine learning introduce the potential for significant advancements in data analysis and pattern recognition. These algorithms leverage quantum parallelism to process large datasets more efficiently and may lead to breakthroughs in fields like drug discovery, finance, and recommendation systems. Quantum algorithms for cryptography play a crucial role in securing communication and data in the quantum era. These algorithms aim to develop cryptographic protocols that are resilient to attacks from quantum computers, ensuring data privacy and integrity in the quantum age. Quantum-resistant encryption algorithms are of particular interest, as they offer security against quantum attacks. Quantum algorithms for artificial intelligence and deep learning explore the use of quantum computers to accelerate machine learning tasks, potentially revolutionizing the field of AI. These algorithms seek to improve the efficiency of training deep neural networks, optimizing parameter searches, and enhancing data preprocessing. Quantum algorithms for optimization offer the potential to solve complex optimization problems with unprecedented efficiency, impacting various industries such as finance, logistics, and energy. These algorithms leverage quantum parallelism to explore solution spaces and find optimal solutions faster than classical counterparts. Quantum algorithms for quantum chemistry aim to simulate the behavior of molecules and materials with high precision. These algorithms have the potential to revolutionize drug discovery, materials science, and chemical engineering by providing insights into molecular interactions and properties. Quantum algorithms for quantum simulations are essential for studying quantum systems that are difficult to simulate classically. These algorithms enable researchers to explore quantum phenomena, such as quantum phase transitions and exotic quantum states, with unprecedented accuracy. Quantum algorithms for finance offer the potential to revolutionize the financial industry by optimizing portfolio management, risk assessment, and algorithmic trading strategies. These algorithms leverage quantum parallelism to process vast amounts of financial data efficiently, potentially providing a competitive edge in financial markets. Quantum algorithms for natural language processing aim to enhance the processing of human language data, enabling more advanced and efficient language understanding and generation tasks. These algorithms have applications in areas such as machine translation, sentiment analysis, and text summarization. Quantum algorithms for network analysis are designed to explore complex network structures and solve problems related to graph theory. These algorithms have applications in various domains, including transportation, social networks, and cybersecurity. Quantum algorithms for machine learning combine the power of quantum computing with machine learning techniques to improve the efficiency of tasks such as data classification, regression, and clustering. These algorithms have the potential to advance fields like healthcare, finance, and image recognition. Quantum algorithms for optimization focus on solving complex optimization problems efficiently, with applications in fields such as logistics, finance, and engineering. These algorithms leverage quantum parallelism to explore solution spaces and find optimal solutions more rapidly than classical methods. Quantum algorithms for cryptography are essential for ensuring the security of communication and data in the quantum era. These algorithms aim to develop cryptographic protocols that are resilient to attacks from quantum computers, protecting data privacy and integrity. Quantum algorithms for artificial intelligence and deep learning seek to accelerate machine learning tasks by leveraging the computational power of quantum computers. These algorithms aim to improve the efficiency of training deep neural networks, optimizing parameter searches, and enhancing data preprocessing. In summary, advanced quantum algorithm design encompasses a diverse array of algorithms that harness the unique properties of quantum systems to solve complex problems efficiently. From Shor's algorithm in cryptography to quantum machine learning and optimization algorithms, these innovations have the potential to revolutionize various fields and reshape the way we approach computational challenges.

Quantum algorithms represent a transformative approach to tackling complex problems that have long stymied classical computers. These algorithms leverage the unique properties of quantum systems to perform computations with an exponential speedup, promising breakthroughs in fields such as cryptography, optimization, and materials science. One of the *most prominent quantum algorithms, Shor's algorithm, has the remarkable ability to factor large numbers exponentially faster than classical methods. This poses a significant threat to classical encryption schemes that rely on the difficulty of factoring large numbers, as Shor's algorithm could potentially break these cryptographic systems. The implications of Shor's algorithm have prompted the development of post-quantum cryptography, which seeks encryption methods that remain secure in the era of quantum computers. Quantum algorithms for optimization are another crucial area of exploration. These algorithms aim to find the best solution among a vast number of possibilities, a task that is vital in various industries, including logistics, finance, and engineering. The Quantum Approximate Optimization Algorithm (QAOA) is a quantum algorithm that shows promise in solving complex optimization problems efficiently. By harnessing quantum parallelism, QAOA explores the solution space more effectively than classical algorithms, making it a valuable tool for industry and research. Quantum machine learning algorithms represent an exciting fusion of quantum computing and machine learning techniques. These algorithms aim to process and analyze large datasets more efficiently, with potential applications in fields like healthcare, finance, and natural language processing. The Quantum Support Vector Machine (QSVM) is an example of a quantum machine learning algorithm designed to improve classification tasks through quantum parallelism. Quantum algorithms for linear algebra play a critical role in various scientific and engineering applications. These algorithms streamline the process of solving linear equations, matrix inversion, and eigenvalue problems, which are fundamental in physics, engineering, and data analysis. The Quantum Phase Estimation (QPE) algorithm, for instance, provides an exponential speedup for finding eigenvalues of unitary operators, making it invaluable for quantum chemistry simulations and quantum cryptography protocols. Quantum algorithms for graph theory and network analysis offer powerful tools for solving complex problems related to networks. These algorithms utilize quantum principles to explore graph structures efficiently, opening up applications in transportation, logistics, and cybersecurity. Quantum algorithms for combinatorial optimization are tailored to address problems involving the selection of the best combination from a finite set of options. They find applications in resource allocation, scheduling, and portfolio optimization, with the Quantum Approximate Optimization Algorithm (QAOA) leading the way in solving such problems efficiently. Quantum algorithms for machine learning have the potential to revolutionize data analysis and pattern recognition. These algorithms harness quantum parallelism to process large datasets more efficiently and may lead to breakthroughs in fields like drug discovery, finance, and recommendation systems. Quantum algorithms for cryptography are essential for ensuring the security of communication and data in the quantum era. They aim to develop cryptographic protocols that are resilient to attacks from quantum computers, safeguarding data privacy and integrity. Quantum-resistant encryption algorithms are of particular interest, as they offer security against quantum attacks. Quantum algorithms for artificial intelligence and deep learning explore the use of quantum computers to accelerate machine learning tasks, potentially revolutionizing the field of AI. These algorithms seek to enhance the efficiency of training deep neural networks, optimizing parameter searches, and improving data preprocessing. Quantum algorithms for network analysis are designed to explore complex network structures and solve problems related to graph theory. These algorithms have applications in various domains, including transportation, social networks, and cybersecurity. Quantum algorithms for machine learning combine the power of quantum computing with machine learning techniques to improve the efficiency of tasks such as data classification, regression, and clustering. These algorithms have the potential to advance fields like healthcare, finance, and image recognition. Quantum algorithms for optimization focus on solving complex optimization problems efficiently, with applications in fields such as logistics, finance, and engineering. These algorithms leverage quantum parallelism to explore solution spaces and find optimal solutions more rapidly than classical methods. Quantum algorithms for cryptography are essential for ensuring the security of communication and data in the quantum era. These algorithms aim to develop cryptographic protocols that are resilient to attacks from quantum computers, protecting data privacy and integrity. Quantum algorithms for artificial intelligence and deep learning seek to accelerate machine learning tasks by leveraging the computational power of quantum computers. These algorithms aim to improve the efficiency of training deep neural networks, optimizing parameter searches, and enhancing data preprocessing. In summary, quantum algorithms for complex problems represent a revolutionary approach to computation, harnessing the capabilities of quantum systems to solve intricate challenges more efficiently than classical computers. From Shor's algorithm in cryptography to quantum machine learning and optimization algorithms, these innovations have the potential to reshape industries and advance our understanding of the world.*

Chapter 4: Quantum Machine Learning in Expert Applications

Advanced quantum machine learning techniques are at the forefront of cutting-edge research, pushing the boundaries of what's possible in the realm of artificial intelligence. These techniques represent a fusion of quantum computing and machine learning, offering unprecedented computational power to tackle complex problems. Quantum machine learning leverages the unique properties of quantum systems, such as superposition and entanglement, to process and analyze data in ways that classical computers cannot match. One of the foundational concepts in quantum machine learning is quantum parallelism, which allows quantum computers to explore multiple solutions simultaneously. This property is harnessed to accelerate tasks such as optimization, pattern recognition, and data analysis. Quantum machine learning models are designed to harness the computational capabilities of quantum computers efficiently. These models range from quantum variants of classical machine learning algorithms to entirely novel approaches that exploit quantum properties. One notable example is the Quantum Support Vector Machine (QSVM), which is a quantum analogue of the classical support vector machine used for classification tasks. QSVM leverages quantum parallelism to explore multiple solutions simultaneously, potentially leading to faster and more accurate classification results. Quantum neural networks represent a fascinating area of exploration within quantum machine learning. These networks utilize quantum circuits to process and learn from data, opening up new possibilities for deep learning in quantum environments. Quantum neural networks can be particularly useful for tasks such as quantum state tomography and quantum error correction. Quantum-enhanced data preprocessing is another crucial aspect of advanced quantum machine learning. Quantum computers can efficiently perform tasks like data dimensionality reduction, feature selection, and data cleaning, improving the quality of input data for subsequent machine learning tasks. Quantum-enhanced optimization algorithms play a vital role in improving the efficiency of training machine learning models. These algorithms leverage quantum parallelism to explore complex optimization landscapes and find optimal parameter settings more rapidly. Quantum-enhanced clustering algorithms aim to identify meaningful patterns and groupings in data by harnessing quantum parallelism and advanced quantum data representations. Quantum generative models, such as quantum Boltzmann machines, offer exciting possibilities for generating data samples that follow complex distributions, with applications in areas like generative art and data augmentation. Quantum reinforcement learning is an emerging field within quantum machine learning, focused on enhancing the capabilities of reinforcement learning algorithms by leveraging quantum computers. These techniques have potential applications in autonomous systems, robotics, and game-playing agents. Quantum machine learning algorithms for natural language processing are designed to process and understand human language more efficiently, with potential applications in areas like language translation, sentiment analysis, and chatbots. Quantum machine learning can also be applied to the field of quantum chemistry, where it has the potential to accelerate the discovery of new molecules and materials. These applications are vital in drug discovery, materials science, and chemical engineering. Quantum machine learning for finance is an area of significant interest, as it can improve the accuracy of financial predictions, risk assessment, and algorithmic trading strategies. These applications have the potential to revolutionize the financial industry. Quantum machine learning for healthcare focuses on improving patient care and disease diagnosis through advanced data analysis techniques. These algorithms can help identify patterns in medical data, optimize treatment plans, and accelerate drug discovery processes. Quantum machine learning techniques can also play a crucial role in optimizing supply chain management and logistics operations, leading to more efficient and cost-effective distribution systems. Quantum machine learning for image and video analysis has applications in fields such as computer vision, object recognition, and video summarization. These techniques can improve the accuracy of image-based tasks and enhance video content analysis. Quantum machine learning for cybersecurity aims to develop advanced intrusion detection systems and threat detection algorithms that can safeguard computer networks more effectively. These techniques leverage quantum parallelism to analyze vast amounts of network data efficiently. Quantum machine learning for recommendation systems enhances personalized recommendations for users in various domains, including e-commerce, content streaming, and social media. These algorithms can better understand user preferences and make more accurate suggestions. Quantum machine learning for optimization problems has applications in various industries, such as logistics, finance, and engineering. These techniques aim to find optimal solutions to complex problems more efficiently than classical methods. Quantum machine learning for data privacy explores methods for secure data analysis while protecting sensitive information. These techniques allow data to be analyzed without exposing *private details, ensuring privacy in healthcare, finance, and other sensitive domains. Quantum machine learning for materials science accelerates the discovery of new materials with desirable properties, leading to advancements in fields like energy storage, electronics, and materials engineering. Quantum machine learning for natural language processing aims to improve the efficiency and accuracy of language-related tasks, such as language translation, sentiment analysis, and text summarization. These techniques have applications in fields like customer support, content generation, and information retrieval. Quantum machine learning for quantum computing focuses on developing algorithms that can harness the power of quantum computers to solve quantum problems efficiently. These algorithms are essential for tasks such as quantum state preparation, quantum error correction, and quantum simulation. Quantum machine learning for quantum error correction plays a crucial role in improving the reliability and stability of quantum computations. These algorithms are vital for scaling up quantum computers to handle complex tasks effectively. Quantum machine learning for quantum simulation enables researchers to explore quantum systems with high precision, leading to advancements in quantum chemistry, materials science, and fundamental physics. These simulations offer insights into quantum phenomena that are difficult to access through classical methods. In summary, advanced quantum machine learning techniques are poised to revolutionize various industries and scientific domains by harnessing the computational power of quantum computers. From quantum-enhanced data preprocessing to quantum reinforcement learning, these techniques offer innovative solutions to complex problems and hold the promise of driving progress in quantum computing and machine learning.*

Quantum machine learning has made significant strides in recent years, attracting attention across various practical expert domains where complex problems demand innovative solutions. One such domain is healthcare, where quantum machine learning has the potential to revolutionize medical research, diagnosis, and treatment planning. Quantum algorithms can process vast datasets more efficiently, leading to improved patient care and disease understanding. In drug discovery, for instance, quantum machine learning can accelerate the identification of potential drug candidates by simulating molecular interactions with high precision. Quantum chemistry simulations provide insights into the behavior of molecules, enabling researchers to design more effective drugs. Another practical expert domain where quantum machine learning shows promise is finance. The financial industry relies heavily on data analysis, risk assessment, and algorithmic trading strategies. Quantum machine learning algorithms can optimize these processes, leading to more accurate financial predictions and enhanced risk management. Quantum-enhanced portfolio optimization can help investors make informed decisions and maximize returns. Quantum machine learning also has applications in the field of logistics and supply chain management. Efficient resource allocation, route optimization, and demand forecasting are essential in this domain. Quantum algorithms can address these challenges by finding optimal solutions to complex logistics problems, reducing operational costs, and improving supply chain efficiency. Quantum machine learning for energy optimization is crucial in the transition to sustainable energy sources. Quantum algorithms can optimize the design and operation of renewable energy systems, leading to more efficient energy production and consumption. In materials science, quantum machine learning accelerates the discovery of new materials with desired properties. Researchers can use quantum simulations to explore the behavior of materials at the quantum level, enabling advancements in electronics, energy storage, and materials engineering. Quantum machine learning is also making strides in cybersecurity. With the increasing complexity of cyber threats, quantum algorithms can enhance intrusion detection systems and threat detection algorithms. These techniques leverage quantum parallelism to analyze vast amounts of network data efficiently, protecting critical systems from cyberattacks. Quantum machine learning for recommendation systems is transforming the way businesses personalize user experiences. E-commerce platforms, content streaming services, and social media networks use quantum-enhanced recommendation algorithms to provide more relevant suggestions to users. In natural language processing, quantum machine learning enhances language-related tasks such as language translation, sentiment analysis, and text summarization. These techniques are essential for applications like customer support, content generation, and information retrieval. Quantum machine learning for optimization problems finds applications in various industries, from logistics and finance to engineering. These algorithms aim to find optimal solutions to complex problems more efficiently than classical methods, leading to cost savings and improved decision-making. Quantum machine learning for data privacy addresses the growing concern about data security and privacy. These techniques allow data to be analyzed while protecting sensitive information, ensuring compliance with data protection regulations. In the field of quantum computing, quantum machine learning plays a critical role in solving quantum problems efficiently. These algorithms are essential for tasks such as quantum state preparation, quantum error correction, and quantum simulation. Quantum machine learning for quantum error correction enhances the reliability and stability of quantum computations. These algorithms are vital for scaling up quantum computers to handle complex tasks effectively. Quantum machine learning for quantum simulation enables researchers to explore quantum systems with high precision, leading to advancements in quantum chemistry, materials science, and fundamental physics. These simulations offer insights into quantum phenomena that are difficult to access through classical methods. In expert domains such as healthcare, finance, logistics, energy, materials science, cybersecurity, recommendation systems, natural language processing, optimization, data privacy, quantum computing, quantum error correction, and quantum simulation, quantum machine learning is proving to be a game-changer. It offers innovative solutions to complex problems, drives progress in quantum computing and machine learning, and holds the promise of transforming industries and scientific research. As quantum machine learning continues to advance, its impact on practical expert domains will undoubtedly grow, ushering in a new era of innovation and discovery.

Chapter 5: Quantum Simulation: Complex Systems and Quantum Chemistry

Quantum simulations have emerged as a powerful tool for exploring and understanding complex systems that are difficult to study using classical computers. These complex systems span a wide range of scientific disciplines, from quantum chemistry and condensed matter physics to materials science and high-energy physics. One of the primary motivations behind quantum simulations is to accurately model quantum systems themselves, providing insights into the behavior of atoms, molecules, and materials at the quantum level. Quantum computers, with their inherent quantum properties, offer the computational resources needed to simulate quantum systems more efficiently than classical computers. Quantum simulations have the potential to revolutionize the field of quantum chemistry by enabling the study of molecular structures, reactions, and properties with unprecedented accuracy. Traditional methods for simulating quantum systems, such as Density Functional Theory (DFT), often suffer from limitations in computational power and accuracy, which quantum simulations can address. Quantum algorithms like the Variational Quantum Eigensolver (VQE) and Quantum Phase Estimation (QPE) hold promise for solving molecular energy and electronic structure problems. In condensed matter physics, quantum simulations are essential for investigating the behavior of materials at the quantum scale. Understanding the properties of materials, such as superconductivity and magnetism, requires detailed quantum modeling, which quantum computers are well-suited to provide. Quantum simulations can help researchers design new materials with tailored properties, advancing fields like materials science and nanotechnology. Complex systems in high-energy physics, such as particle collisions and quantum field theories, are challenging to study due to their intricate mathematical descriptions. Quantum simulations offer a path to tackle these problems by providing a platform to simulate and analyze these systems accurately. Simulating quantum field theories and particle interactions can lead to deeper insights into the fundamental laws of the universe. Quantum simulations are also valuable for studying the behavior of exotic matter under extreme conditions, such as in neutron stars or black holes. These simulations can help researchers probe the properties of matter under extreme gravitational forces, shedding light on the nature of spacetime itself. Quantum simulations are not limited to physics; they have applications in other scientific domains as well. For instance, in biology, simulating the behavior of molecules, proteins, and enzymes at the quantum level can aid drug discovery and the development of new therapies. Quantum simulations enable researchers to understand the quantum properties of biomolecules, leading to advancements in personalized medicine. In the field of optimization, quantum simulations can tackle complex optimization problems by leveraging quantum parallelism. Problems like portfolio optimization, logistics, and scheduling can benefit from quantum algorithms like the Quantum Approximate Optimization Algorithm (QAOA). Quantum simulations are also valuable in finance, where they can be applied to risk assessment, option pricing, and portfolio management, leading to more informed investment strategies. Simulating quantum systems on quantum hardware, however, presents its own set of challenges. Quantum computers are susceptible to errors due to factors like decoherence and gate imperfections. Quantum error correction codes, such as the surface code, are essential for mitigating errors and ensuring the reliability of quantum simulations. Efforts to develop fault-tolerant quantum computing systems are ongoing, with the goal of achieving accurate and stable quantum simulations. Quantum simulations also require quantum software development tools and programming languages tailored to the unique capabilities of quantum hardware. Quantum programming languages like Qiskit, Quipper, and Cirq provide the means to design, implement, and execute quantum algorithms for simulations. Quantum software development tools, including quantum compilers and simulators, facilitate the translation of quantum algorithms into executable code on quantum computers. Quantum simulations can be run on various types of quantum hardware, including gate-based quantum computers, adiabatic quantum computers, and quantum annealers. Each of these platforms has its strengths and limitations, and the choice of hardware depends on the specific simulation requirements. Quantum annealers, for instance, are well-suited for solving optimization problems, while gate-based quantum computers offer more flexibility in simulating general quantum systems. Quantum simulations also benefit from advancements in quantum hardware technologies, such as superconducting qubits and trapped ions, which enable the creation of larger and more capable quantum systems. Simulating complex quantum systems often involves the preparation of quantum states, the application of quantum gates, and the measurement of quantum observables. Quantum algorithms, including the Quantum Phase Estimation (QPE) algorithm, play a crucial role in these simulations by efficiently estimating eigenvalues and extracting valuable information about the system's behavior. Quantum simulations have practical applications beyond fundamental research. For instance, simulating the behavior of new materials can lead to the discovery of novel materials with specific properties, revolutionizing industries like electronics, energy storage, and manufacturing. Quantum simulations can also improve the efficiency of chemical reactions, leading to advancements in areas like catalysis and green chemistry. Furthermore, quantum simulations have the potential to optimize supply chain logistics, reduce energy consumption, and enhance financial modeling. In summary, quantum simulations represent a groundbreaking approach to understanding and solving complex problems across various scientific and practical domains. These simulations leverage the computational power of quantum computers to model quantum systems accurately and efficiently, opening up new frontiers in research and applications. As quantum hardware continues to advance and quantum algorithms evolve, the impact of quantum simulations on science and industry is expected to grow significantly, paving the way for transformative discoveries and innovations. Quantum chemistry, a branch of quantum mechanics, has been instrumental in unraveling the behavior of atoms and molecules at the fundamental level. Its applications extend across various scientific disciplines, offering insights into chemical reactions, molecular structures, and electronic properties. Quantum chemistry provides a robust framework for simulating molecular systems accurately, making it indispensable in fields such as drug discovery, materials science, and catalysis. One of the primary applications of quantum chemistry lies in predicting molecular properties and behaviors with a high degree of precision. Quantum simulations enable researchers to explore potential energy surfaces and reaction pathways, facilitating the discovery of new chemical compounds and reactions. In drug discovery, quantum chemistry plays a pivotal role by assisting in the design of pharmaceutical molecules with desired properties. These simulations provide insights into the binding affinities between drugs and their target proteins, helping pharmaceutical companies develop more effective treatments. Quantum chemistry also contributes to materials science by elucidating the electronic structure and properties of materials. Understanding materials at the quantum level is crucial for optimizing their performance in applications ranging from semiconductors and superconductors to catalysts and nanomaterials. Quantum chemistry simulations guide the development of advanced materials with tailored properties. Furthermore, catalysis, the process of accelerating chemical reactions, benefits from quantum chemistry's ability to predict reaction mechanisms and identify efficient catalysts. This knowledge is vital for improving industrial processes in fields like petrochemicals and environmental remediation. Quantum chemistry is essential for computational chemistry, where it complements experimental work by providing detailed insights into molecular interactions. The accuracy of quantum chemical calculations is continually improving, enabling researchers to tackle increasingly complex chemical systems. Quantum chemistry also plays a significant role in quantum computing, where it serves as a benchmark for evaluating the capabilities of quantum algorithms and hardware. Quantum computers are expected to outperform classical computers in simulating quantum systems, a development that promises to revolutionize quantum chemistry itself. One of the most notable quantum algorithms for quantum chemistry is the Variational Quantum Eigensolver (VQE), which aims to find the ground state energy of a molecular system efficiently. VQE leverages the quantum parallelism of quantum computers to explore multiple molecular configurations simultaneously, offering a potential advantage over classical *algorithms. Another quantum algorithm, Quantum Phase Estimation (QPE), enables precise determination of molecular eigenvalues and facilitates the simulation of molecular dynamics. Quantum chemistry's potential extends to solving complex problems in quantum materials and condensed matter physics. The behavior of electrons in solid-state materials poses significant challenges due to the intricate interactions among particles. Quantum simulations can shed light on phenomena such as superconductivity, magnetism, and topological phases of matter. By modeling electronic structures with high fidelity, quantum chemistry contributes to advancements in electronic devices, novel materials, and energy technologies. Beyond its applications in traditional chemistry and materials science, quantum chemistry has implications for quantum cryptography. Quantum key distribution (QKD) protocols rely on the principles of quantum mechanics to secure communication channels. The security of QKD protocols is grounded in the fundamental properties of quantum particles, making them resistant to eavesdropping. Quantum chemistry's understanding of quantum states and entanglement is crucial for advancing quantum cryptography and ensuring the confidentiality of sensitive information. Quantum chemistry's reach extends to quantum algorithms for machine learning. Quantum machine learning models leverage quantum computing's capabilities to process and analyze large datasets efficiently. These models have applications in quantum chemistry for predicting molecular properties and optimizing chemical reactions. Quantum machine learning can expedite the drug discovery process by identifying promising drug candidates and simulating their interactions with biological systems. Furthermore, quantum chemistry has applications in quantum-enhanced optimization. Quantum algorithms like the Quantum Approximate Optimization Algorithm (QAOA) are designed to solve complex optimization problems efficiently. In fields such as logistics, finance, and engineering, quantum-enhanced optimization can lead to more cost-effective and resource-efficient solutions. Quantum chemistry's accurate modeling of molecular systems is pivotal in quantum-enhanced optimization, where it helps define objective functions and constraints. Quantum chemistry has made significant strides in quantum computing, with the potential to revolutionize various scientific and industrial domains. As quantum hardware continues to advance and quantum algorithms evolve, quantum chemistry will play an increasingly critical role in unlocking new frontiers in chemistry, materials science, cryptography, machine learning, and optimization. Its applications are not confined to the laboratory; they extend to solving real-world problems that have a profound impact on society and technology. In the quest for advanced materials, novel drugs, secure communication, and efficient resource allocation, quantum chemistry stands as a formidable tool, offering unparalleled insights into the quantum world and its applications beyond.*

Chapter 6: Fault-Tolerant Quantum Computing and Error Correction

Error correction is a crucial aspect of quantum computing, as quantum systems are inherently susceptible to noise and errors due to their fragile quantum states. To achieve reliable and fault-tolerant quantum computations, advanced error correction schemes are essential. Classical error correction techniques, which involve redundant encoding of information, are adapted for quantum systems to combat noise and errors. Quantum error correction codes, such as the surface code and the Steane code, are fundamental to protecting quantum information. The surface code, for instance, encodes qubits in a two-dimensional lattice, allowing for the detection and correction of errors by measuring neighboring qubits. These codes form the foundation of quantum error correction, but they are not without challenges. One of the key challenges in quantum error correction is the need for physical qubits to be more reliable than logical qubits. Logical qubits are encoded using multiple physical qubits, and error correction relies on the redundancy and measurement of these physical qubits. Quantum error correction schemes must address the issue of physical qubit reliability to ensure the overall fault tolerance of quantum computations. The implementation of quantum error correction also requires specialized hardware and software to perform error detection and correction operations. Quantum error correction codes are designed to detect and correct errors by encoding quantum information in a way that allows for error syndromes to be measured without disturbing the encoded information. Error syndromes are patterns of errors that indicate the presence and location of errors in the quantum state. Once error syndromes are measured, quantum error correction algorithms determine how to correct the errors to recover the original quantum information. Quantum error correction goes beyond detecting and correcting single-qubit errors. It also addresses two-qubit errors, correlated errors, and coherent errors that affect multiple qubits simultaneously. Addressing these complex error scenarios is essential for achieving high fault tolerance in quantum computing. Fault tolerance is a critical metric for assessing the effectiveness of quantum error correction schemes. A fault-tolerant quantum computer can perform quantum computations reliably, even in the presence of a certain level of noise and errors. The threshold theorem sets the theoretical limit for fault-tolerant quantum computation. It specifies that if the error rate per physical qubit is below a certain threshold, fault-tolerant quantum computing becomes feasible. Developing quantum error correction schemes that approach or surpass this threshold is a significant research goal in quantum computing. Quantum error correction schemes are also categorized based on their approach to error correction, such as quantum codes, quantum stabilizer codes, and quantum LDPC codes. Quantum codes use additional qubits to encode information in a manner that allows errors to be detected and corrected. Quantum stabilizer codes, on the other hand, use stabilizer generators to define the code space and correct errors. Quantum LDPC codes (Low-Density Parity-Check codes) are another class of quantum error correction codes that are known for their efficiency and low error threshold. These codes are characterized by sparse parity-check matrices, making them suitable for certain quantum error correction applications. Quantum error correction is not limited to theoretical research; it is a practical necessity for building scalable and reliable quantum computers. Several quantum computing platforms, such as superconducting qubits and trapped ions, have implemented quantum error correction codes to mitigate the impact of errors. For example, some quantum processors use concatenated codes, which combine multiple layers of error correction codes to achieve higher fault tolerance. Implementing quantum error correction on real quantum hardware requires addressing challenges related to qubit connectivity, gate fidelities, and measurement errors. Moreover, quantum error correction is intertwined with quantum fault tolerance, as achieving high fault tolerance depends on the choice of quantum error correction codes and their efficient implementation. Quantum fault tolerance also considers the overhead associated with encoding, detecting, and correcting errors, which can be substantial. As quantum hardware advances, researchers continue to explore new quantum error correction codes and techniques to improve fault tolerance. These advancements aim to bring practical, fault-tolerant quantum computing closer to reality. Quantum error correction codes have far-reaching implications beyond quantum computing. They are also essential in quantum communication, where quantum information must be transmitted reliably over long distances. Quantum key distribution (QKD) protocols, for instance, rely on error correction to secure quantum communication channels against eavesdropping. Quantum error correction is an interdisciplinary field that bridges quantum information theory, quantum coding theory, and quantum hardware engineering. It requires a deep understanding of quantum mechanics and the mathematical principles underlying error correction codes. Quantum error correction research is closely linked to quantum algorithms, as quantum computers are designed to solve complex problems that classical computers cannot. Quantum algorithms, in turn, benefit from the error correction schemes that ensure the reliability of quantum computations. In summary, advanced error correction schemes are at the heart of quantum computing's quest for fault tolerance and scalability. These schemes leverage the principles of quantum mechanics to detect and correct errors, ensuring the integrity of quantum information. As quantum hardware continues to evolve, so too will the sophistication of quantum error correction techniques, bringing us closer to practical, fault-tolerant quantum computing and secure quantum communication. Fault tolerance is a paramount concern in quantum computing and quantum information processing, as quantum systems are inherently susceptible to errors and noise that can compromise the accuracy and reliability of quantum computations. Achieving fault tolerance in quantum systems is essential for realizing the full potential of quantum technology and harnessing its computational power for practical applications. Quantum fault tolerance ensures that quantum computers can perform reliable computations even in the presence of errors, which is critical for solving complex problems that classical computers cannot. The concept of fault tolerance is rooted in the principles of quantum error correction, which involves encoding quantum information in a way that allows errors to be detected and corrected. Quantum error correction codes, such as the surface code and the Steane code, play a pivotal role in protecting quantum information from errors and noise. These codes introduce redundancy by encoding logical qubits using multiple physical qubits, allowing for the detection and correction of errors through measurements of the physical qubits. The surface code, for example, encodes qubits in a two-dimensional lattice, with each qubit entangled with its neighbors. Errors can be detected by measuring the stabilizer generators, which are operators *that check for deviations from the expected quantum states. Quantum fault tolerance relies on the fault-tolerant operations of quantum error correction codes, which are capable of recovering from errors without introducing additional errors. Achieving fault tolerance requires quantum computers to meet certain error rate thresholds, as specified by the threshold theorem. This theorem establishes that if the error rate per physical qubit is below a certain threshold, fault-tolerant quantum computation becomes feasible. Thus, reducing the error rate of quantum hardware is a central objective in the pursuit of fault tolerance. Several approaches are employed to lower the error rate, including improving the physical qubit quality, implementing error-correcting codes, and developing quantum hardware with increased qubit connectivity and gate fidelities. Quantum error correction, while foundational to fault tolerance, introduces significant overhead in terms of the number of physical qubits required for encoding, the additional gates needed for error correction, and the increased complexity of quantum algorithms. Addressing these challenges is crucial to achieving practical fault tolerance. Quantum fault tolerance goes beyond mitigating single-qubit errors; it also addresses two-qubit errors, correlated errors, and coherent errors that affect multiple qubits simultaneously. These error scenarios are complex and require advanced error correction techniques to maintain computational integrity. Coherent errors, in particular, pose a significant challenge, as they can affect multiple qubits coherently and propagate throughout the quantum system. Quantum fault tolerance strategies must account for these coherent errors to ensure fault-tolerant quantum computations. Quantum fault tolerance also extends to fault-tolerant quantum gates, which are essential for performing quantum computations reliably. Fault-tolerant gates are designed to operate on encoded qubits while preserving the error-corrected state. Implementing fault-tolerant gates involves a careful sequence of quantum operations that correct errors and maintain the logical quantum state. These gates are crucial for building fault-tolerant quantum circuits that can execute complex quantum algorithms. Quantum fault tolerance has practical implications for quantum hardware development, as achieving fault-tolerant quantum computation necessitates advancements in qubit quality, gate fidelities, and error rates. Superconducting qubits, trapped ions, and other quantum technologies are actively pursuing fault tolerance by improving the performance and reliability of quantum hardware. Quantum error correction codes, tailored to specific hardware platforms, are employed to address the unique error characteristics of each technology. Quantum fault tolerance is closely linked to the broader field of quantum error correction, which encompasses the development of new quantum codes, error-correcting algorithms, and quantum fault-tolerant architectures. Efforts to achieve fault tolerance are driven by the potential of quantum computing to revolutionize fields such as cryptography, optimization, materials science, and drug discovery. Quantum computers hold the promise of solving complex problems at speeds unimaginable to classical computers, provided they can maintain their computational integrity in the presence of errors. Quantum fault tolerance is instrumental in realizing this potential and making quantum computing a practical tool for scientific research and industrial applications. As quantum hardware continues to evolve and improve, the threshold for fault-tolerant quantum computation is expected to be reached, ushering in a new era of reliable and powerful quantum computing.*

Chapter 7: Quantum Hardware Innovations and Scalability

Quantum hardware is at the forefront of technological innovation, with continuous advancements that hold the promise of transforming computing and communication. These cutting-edge developments in quantum hardware are shaping the future of quantum technology, with the potential to revolutionize industries and scientific research. One of the most exciting developments in quantum hardware is the pursuit of scalable and fault-tolerant quantum processors. Quantum processors are the heart of quantum computers, and efforts are underway to increase the number of qubits while maintaining low error rates. Superconducting qubits and trapped ions are two leading approaches, with each offering unique advantages. Superconducting qubits are integrated into chip-based architectures, allowing for easy scalability and compatibility with existing semiconductor technology. Trapped ions, on the other hand, benefit from exceptionally long qubit coherence times, making them well-suited for error-corrected quantum computations. In addition to qubit quantity and quality, quantum hardware is advancing in terms of qubit connectivity. Enhanced qubit connectivity enables more complex quantum algorithms and simulations. Advanced hardware designs are being explored to enable long-range qubit interactions, facilitating the creation of entangled states across multiple qubits. This is essential for solving complex problems in fields like materials science and quantum chemistry. Another cutting-edge development in quantum hardware is the integration of quantum processors with classical computing resources. Quantum co-processors, which combine classical and quantum hardware, aim to leverage the strengths of both paradigms for enhanced computational capabilities. Quantum co-processors can be used for quantum simulations, optimization problems, and machine learning tasks, providing a bridge between classical and quantum computing. Quantum hardware is also making strides in enhancing gate fidelities and reducing noise levels. This is achieved through better qubit control, error correction techniques, and innovative qubit designs. Higher gate fidelities are crucial for executing quantum algorithms with fewer errors, ultimately leading to more accurate results. Moreover, quantum hardware development is closely linked to quantum error correction, as achieving fault-tolerant quantum computation relies on low error rates and reliable quantum gates. A significant challenge in quantum hardware is minimizing decoherence, which refers to the loss of quantum information due to interactions with the environment. Decoherence reduces the lifetime of quantum states and limits the duration of quantum computations. Researchers are exploring various techniques, such as error-correcting codes and dynamic error correction, to mitigate the effects of decoherence and extend the coherence times of qubits. Another exciting development is the exploration of topological qubits, which are inherently robust against certain types of errors. Topological qubits are expected to play a vital role in the realization of fault-tolerant quantum computers. Quantum hardware is not limited to quantum processors alone; it also encompasses quantum memory and quantum interconnects. Quantum memory enables the storage and retrieval of quantum states, which is essential for quantum communication and distributed quantum computing. Quantum interconnects facilitate the transfer of quantum information between quantum processors, enabling large-scale quantum networks and quantum cloud computing. The development of quantum memory and interconnects is critical for building a quantum internet and enabling secure quantum communication protocols. Quantum hardware is advancing rapidly, driven by collaboration between academia and industry. Quantum computing companies are competing to build increasingly powerful quantum processors and make them accessible through cloud-based platforms. These platforms allow researchers and businesses to experiment with quantum hardware, accelerating the development of quantum algorithms and applications. Moreover, quantum hardware is not confined to research laboratories; it is increasingly being used in real-world applications. Quantum sensors, for instance, leverage quantum technology to achieve unprecedented levels of sensitivity and accuracy in fields like geophysics, medical imaging, and navigation. Quantum-enhanced sensors can detect subtle changes in physical properties, opening up new possibilities for scientific discovery and industrial applications. Quantum hardware also holds the promise of advancing cryptography through quantum-resistant encryption algorithms. As quantum computers become more powerful, they pose a threat to classical cryptographic systems, making the development of quantum-resistant cryptography a pressing concern. Quantum hardware can contribute to the development and deployment of encryption methods that can withstand attacks from quantum adversaries. In summary, cutting-edge developments in quantum hardware are driving the evolution of quantum technology. The pursuit of scalable and fault-tolerant quantum processors, enhanced qubit connectivity, and integration with classical computing resources are key milestones in the field. Reducing decoherence, exploring topological qubits, and advancing quantum memory and interconnects are essential for building practical quantum systems. Quantum hardware is no longer a theoretical concept; it is becoming a tangible and transformative force in the worlds of computing, communication, and sensing. As quantum hardware continues to advance, its impact on science, industry, and society will become increasingly profound, ushering in a new era of technological innovation. Scalability is a fundamental challenge in quantum computing, as it involves increasing the size and complexity of quantum systems to solve more substantial problems. The scalability challenge arises due to the delicate nature of quantum information and the susceptibility of quantum states to noise and errors. One of the primary issues in scaling quantum computers is maintaining the coherence of qubits as the system grows in size. Qubits are highly sensitive to their environment, and as more qubits are added to a quantum computer, the probability of errors and decoherence also increases. This sensitivity poses a significant hurdle to achieving the scalability required for practical quantum computing applications. Efforts to address this challenge include developing error-correction codes that can detect and correct errors in large-scale quantum systems. These codes introduce redundancy by encoding logical qubits using multiple physical qubits, allowing for the detection and correction of errors through measurements of the physical qubits. While error correction is a crucial step towards scalability, it comes with its own set of challenges, such as the overhead introduced by encoding and the need for high-fidelity gates. Another aspect of scalability in quantum computing is the connectivity between qubits. As the number of qubits grows, ensuring that they can interact with each other becomes increasingly challenging. Long-range qubit connectivity is essential for implementing quantum algorithms efficiently. Advanced hardware designs are exploring ways to achieve long-range interactions, such as coupling qubits through intermediate qubits or using superconducting resonators. These developments aim to enhance the connectivity of large-scale quantum processors. Furthermore, quantum computing platforms are transitioning from small-scale, isolated quantum devices to interconnected quantum networks. Quantum network scalability relies on quantum repeaters and quantum routers that can transmit quantum information over long distances. Quantum repeaters are essential for achieving secure quantum communication and building a quantum internet. The scalability of quantum algorithms is another critical aspect of quantum computing. Quantum algorithms are designed to exploit the computational advantages of quantum systems, but they must be adaptable to larger quantum processors. Developing scalable quantum algorithms requires rethinking algorithms to take advantage of increased qubit resources efficiently. Quantum machine learning, quantum optimization, and quantum simulations are fields where scalable quantum algorithms have the potential to deliver significant advantages. Hybrid quantum-classical algorithms are also being developed to solve complex problems by combining classical and quantum resources, making them adaptable to large-scale quantum processors. Quantum error correction is central to addressing scalability challenges in quantum computing. To achieve scalable and fault-tolerant quantum computation, error correction must be integrated seamlessly into quantum algorithms and hardware. Quantum fault tolerance relies on the threshold theorem, which sets an error rate threshold below which fault-tolerant quantum computation becomes feasible. Reducing error rates through hardware improvements and innovative error-correction techniques is essential for achieving scalability. Quantum supremacy, the point at which quantum computers can outperform classical computers on certain tasks, is a significant milestone on the path to scalability. Achieving quantum supremacy requires demonstrating that quantum processors can perform calculations that are practically infeasible for classical computers to replicate. Google's 53-qubit quantum processor, Sycamore, claimed to achieve quantum supremacy by performing a specific task faster than the world's most advanced supercomputers. While this achievement marked a significant step towards scalability, challenges remain in scaling quantum processors to handle more practical and diverse applications. Quantum annealers, such as those developed by D-Wave Systems, are another avenue for quantum computing scalability. These specialized quantum processors are designed for optimization problems and have *shown promise in various industries, including finance, logistics, and drug discovery. Quantum annealers leverage quantum effects to explore large solution spaces efficiently, making them valuable tools for scalable optimization. The development of quantum programming languages and quantum software development tools is critical for scaling quantum computing. These tools aim to simplify the process of programming and optimizing quantum algorithms, allowing researchers and developers to harness the power of large-scale quantum processors effectively. Quantum cloud computing platforms, offered by companies like IBM, Amazon, and Microsoft, provide access to quantum hardware and software resources in the cloud, enabling scalability for a broader user base. The scalability of quantum computing also has implications for quantum cryptography and quantum-secure communication. As quantum computers grow in power, classical encryption methods become vulnerable to attacks. Quantum-resistant cryptographic techniques, which are designed to withstand attacks from quantum adversaries, are crucial for scalable and secure communication. In summary, scalability is a critical challenge in quantum computing, encompassing various aspects, including qubit coherence, connectivity, algorithms, error correction, and software development. Efforts to address these challenges are ongoing, driven by collaboration between academia and industry. While quantum supremacy and specialized quantum processors represent significant milestones, the journey towards scalable and practical quantum computing continues. Overcoming scalability challenges will unlock the full potential of quantum technology, enabling breakthroughs in fields such as materials science, cryptography, optimization, and drug discovery. As quantum processors continue to scale, their impact on science, industry, and society is expected to be transformative, ushering in a new era of computational capabilities and technological innovation.*

Chapter 8: Quantum Programming Paradigms and Languages

Advanced quantum programming paradigms are paving the way for harnessing the immense computational power of quantum computers more effectively. These paradigms extend beyond the conventional quantum circuit model and offer new ways to express quantum algorithms. One such paradigm is quantum machine learning, where quantum computers are used to enhance machine learning algorithms and solve complex problems. Quantum machine learning combines the principles of quantum mechanics with classical machine learning techniques to achieve tasks that are beyond the capabilities of classical computers. Quantum machine learning algorithms leverage quantum properties such as superposition and entanglement to perform tasks like data classification, clustering, and optimization more efficiently. One of the groundbreaking algorithms in this paradigm is the Quantum Support Vector Machine (QSVM), which can classify data with exponential speedup over classical algorithms. Quantum machine learning is expected to find applications in fields like finance, drug discovery, and image analysis, where large-scale data processing is a challenge. Another advanced quantum programming paradigm is quantum annealing, which is particularly suited for solving optimization problems. Quantum annealers, like those developed by D-Wave Systems, use quantum annealing to explore complex solution spaces and find optimal solutions faster than classical optimization methods. Quantum annealing offers a hybrid approach, where quantum hardware can be integrated with classical algorithms to achieve better results. Quantum annealers have applications in logistics, finance, and materials science, where optimization is a critical component. Quantum programming paradigms also include quantum algorithms for solving specific problems, such as Shor's algorithm for factoring large numbers and Grover's algorithm for searching unsorted databases. These algorithms demonstrate the potential for exponential speedup in solving certain problems compared to classical algorithms. Advanced quantum programming languages and development environments are essential for realizing the full potential of quantum computing. Languages like Qiskit, Cirq, and Quipper provide high-level abstractions for expressing quantum algorithms, making quantum programming more accessible to researchers and developers. These languages come with libraries for quantum circuit construction, optimization, and simulation. Quantum software development tools, such as Qiskit Aqua and PennyLane, focus on quantum applications like chemistry simulations and quantum machine learning. They aim to streamline the development of quantum algorithms for specific domains. Quantum programming languages and tools also enable the exploration of hybrid quantum-classical algorithms, where quantum and classical processors work together to solve complex problems. Hybrid quantum-classical algorithms leverage the strengths of both classical and quantum computing, allowing quantum processors to focus on tasks where they provide the most advantage. In the realm of quantum programming paradigms, quantum-inspired algorithms deserve mention. These algorithms are designed to run on classical computers but are inspired by quantum principles. Quantum-inspired algorithms mimic quantum algorithms' behavior, such as quantum walks or quantum annealing, to achieve performance improvements in classical computing. They are a bridge between classical and quantum computing, offering enhanced solutions for problems that do not yet benefit from full quantum computation. Quantum-inspired algorithms are particularly relevant for situations where access to quantum hardware is limited. As quantum programming paradigms continue to evolve, researchers are exploring ways to optimize quantum algorithms for practical applications. Efforts are underway to develop quantum compilers that can translate high-level quantum code into efficient quantum circuit implementations. Quantum compilers aim to reduce the overhead associated with error correction and improve the overall performance of quantum algorithms. Additionally, quantum programming paradigms are essential for addressing the challenges of error correction and fault tolerance in quantum computing. Quantum error correction codes, such as the surface code, are a fundamental aspect of quantum programming for fault tolerance. These codes introduce redundancy to protect quantum information from errors and enable reliable quantum computations. Advanced quantum programming paradigms play a pivotal role in integrating error correction techniques seamlessly into quantum algorithms and circuits. Quantum programming also extends to the field of quantum simulators, which are specialized quantum devices or software tools that emulate the behavior of quantum systems. Quantum simulators are invaluable for testing and verifying quantum algorithms before running them on actual quantum hardware. They help researchers understand the behavior of quantum systems, explore the effects of noise and errors, and optimize algorithms for real-world quantum processors. Quantum simulators can simulate quantum chemistry experiments, quantum materials, and many-body quantum systems, providing valuable insights for scientific research. In summary, advanced quantum programming paradigms encompass a diverse range of approaches and tools for harnessing the power of quantum computing. From quantum machine learning to quantum annealing and quantum-inspired algorithms, these paradigms offer new ways to express quantum algorithms and solve complex problems more efficiently. Quantum programming languages, development environments, and software tools are crucial for enabling researchers and developers to work with quantum technology effectively. As quantum hardware continues to advance, the role of quantum programming in realizing the potential of quantum computing becomes increasingly significant. Quantum programming paradigms are at the forefront of the quantum revolution, shaping the future of computation, simulation, and optimization. Quantum software development languages and tools are the foundation of quantum computing, enabling researchers and developers to write, test, and optimize quantum algorithms. These specialized languages and tools are essential because quantum computing operates under unique principles distinct from classical computing. One of the pioneering quantum software development languages is Qiskit, developed by IBM. Qiskit is an open-source framework that provides a high-level, Python-based interface for programming quantum circuits and conducting quantum experiments. It offers a user-friendly environment for designing quantum algorithms and simulating their behavior on classical computers. With Qiskit, users can create quantum circuits by defining quantum gates and measurements, and then execute these circuits on real quantum hardware or simulators. Another popular quantum software development language is Cirq, developed by Google. Cirq is an open-source framework that focuses on quantum programming for noisy intermediate-scale quantum (NISQ) devices. It allows researchers to write quantum circuits using Python and provides tools for running *circuits on quantum processors and simulators. Cirq's design emphasizes flexibility and control, making it suitable for exploring quantum algorithms, optimizing quantum gates, and working with NISQ devices. Quipper is another quantum software development language that offers a high-level language for quantum programming. Developed by Microsoft Research and the University of Oxford, Quipper is designed to facilitate the development of quantum algorithms and their mapping to physical quantum hardware. It supports classical and quantum components, allowing for hybrid quantum-classical computations. While these languages offer diverse approaches to quantum programming, they all share the goal of making quantum computing accessible to a broader audience. Quantum software development tools extend beyond programming languages and include comprehensive frameworks for quantum application development. One such tool is Qiskit Aqua, a quantum software development kit (SDK) built on top of Qiskit. Qiskit Aqua is tailored for solving problems in quantum chemistry, optimization, and machine learning. It provides a collection of quantum algorithms and domain-specific libraries, making it easier to develop quantum applications in these fields. Aqua's modular architecture allows developers to combine different components to create custom quantum algorithms and applications. Another quantum software development tool is PennyLane, developed by Xanadu, which focuses on quantum machine learning and quantum optimization. PennyLane is designed to integrate quantum computing into existing machine learning and optimization workflows. It offers a Python-based interface for quantum circuit construction, evaluation, and optimization, allowing researchers to combine quantum and classical processing seamlessly. PennyLane supports various quantum hardware platforms and simulators, making it versatile for quantum application development. Quantum software development languages and tools also play a crucial role in addressing the challenges of quantum error correction. Quantum error correction is essential for achieving fault-tolerant quantum computation, but it introduces significant complexity into quantum algorithms and circuits. Quantum programming languages like Qiskit and Cirq provide support for error-correcting codes and operations, making it easier for developers to work with fault-tolerant quantum systems. Furthermore, quantum software development tools offer specialized modules for error correction, allowing researchers to experiment with error-corrected quantum algorithms. Quantum software development is closely linked to quantum software development kits (SDKs), which are collections of libraries, tools, and documentation for quantum application development. SDKs provide a comprehensive ecosystem for quantum programmers, enabling them to access quantum hardware, run experiments, and analyze results effectively. For example, IBM's Qiskit is not only a quantum programming language but also a quantum software development kit that includes a range of components for quantum computing. These components include Qiskit Terra for circuit construction, Qiskit Aer for quantum simulation, Qiskit Aqua for quantum applications, and Qiskit Ignis for quantum error correction. Qiskit's modular architecture makes it easy for developers to choose the components they need for their specific quantum tasks. Similarly, Google's Cirq provides a full-stack quantum development framework, including libraries for quantum circuit definition, noise modeling, and quantum hardware access. Cirq's versatility allows developers to create quantum circuits and simulations tailored to their research or application needs. Quantum software development tools also enable developers to work with quantum cloud computing platforms offered by major technology companies. These platforms provide cloud-based access to quantum hardware, allowing researchers and developers to experiment with real quantum processors. IBM Quantum Experience, for instance, offers cloud-based access to IBM's quantum devices through the Qiskit platform. Amazon Braket and Microsoft Quantum Development Kit provide similar cloud-based quantum computing services, further democratizing quantum research and development. Quantum software development languages and tools are crucial for realizing the full potential of quantum computing. They empower researchers and developers to explore quantum algorithms, experiment with quantum hardware, and create quantum applications. Quantum programming languages like Qiskit, Cirq, and Quipper offer user-friendly interfaces for expressing quantum algorithms, while quantum software development tools like Aqua and PennyLane focus on specific quantum application domains. SDKs like Qiskit and Cirq provide comprehensive ecosystems for quantum programming and experimentation. With the growing availability of quantum cloud computing platforms, quantum software development is becoming increasingly accessible, enabling a broader community to engage in quantum research and innovation. In summary, quantum software development languages and tools are indispensable for the development of quantum algorithms and applications. They bridge the gap between quantum theory and practical quantum programming, making it possible for researchers and developers to harness the power of quantum computing for diverse fields such as quantum chemistry, optimization, machine learning, and more.*

Chapter 9: Quantum Cryptography and Secure Quantum Communication

Advanced quantum cryptography protocols are at the forefront of securing sensitive information in an era of increasingly powerful quantum computers. These protocols leverage the unique properties of quantum mechanics to provide a level of security that classical cryptographic methods cannot match. Quantum cryptography, also known as quantum key distribution (QKD), is the science of using quantum properties to establish secure communication channels. One of the most well-known quantum cryptography protocols is the BBM92 protocol, developed by Charles Bennett and Gilles Brassard in 1992. The BBM92 protocol is based on the fundamental principle of quantum entanglement, where two particles, such as photons, become correlated in such a way that the state of one is dependent on the state of the other, regardless of the distance between them. In the BBM92 protocol, a sender (Alice) prepares a series of entangled photon pairs and sends one photon from each pair to a receiver (Bob) over a quantum channel. Alice randomly encodes each photon in one of four polarization states: horizontal, vertical, diagonal, or anti-diagonal. Bob measures the polarization of each received photon using a randomly chosen measurement basis. The correlation between Alice's encoding and Bob's measurements allows them to establish a secret key while detecting any eavesdropping attempts by an adversary (Eve). The security of the BBM92 protocol relies on the principles of quantum mechanics, specifically the no-cloning theorem, which states that an arbitrary unknown quantum state cannot be copied exactly. This property prevents Eve from making copies of the transmitted photons without being detected. Another prominent quantum cryptography protocol is the E91 protocol, proposed by Artur Ekert in 1991. The E91 protocol is based on the phenomenon of quantum entanglement and is often referred to as "Einstein-Podolsky-Rosen (EPR) entanglement." In the E91 protocol, two distant parties, Alice and Bob, each generate pairs of entangled particles (usually photons) with correlated polarizations. They send one particle from each pair to a central authority known as Charlie. Charlie performs measurements on the received particles and publicly announces the measurement bases he used but not the measurement outcomes. Alice and Bob then compare their respective measurement bases and retain only the measurement outcomes that correspond to matching bases. The remaining measurement outcomes form their secret key, which is secure against eavesdropping by an adversary. The E91 protocol has the advantage of allowing secure key distribution between two parties even in the presence of a fully cooperative adversary. Quantum cryptography protocols are not limited to key distribution but also encompass secure communication using quantum channels. One such protocol is the quantum teleportation protocol, which enables the transmission of an arbitrary quantum state from one location to another using entanglement and classical communication. Quantum teleportation can be used to establish secure communication channels, as any attempt by an eavesdropper to intercept the transmitted quantum state would disrupt the entanglement and be detectable by the sender and receiver. In addition to key distribution and secure communication, quantum cryptography protocols can also provide security for other cryptographic tasks, such as coin tossing and oblivious transfer. Coin tossing is a cryptographic primitive where two parties want to generate a random bit such that neither party can predict the outcome, and neither can cheat. Quantum coin tossing protocols leverage quantum properties, such as the uncertainty principle, to achieve this level of security. Oblivious transfer is a cryptographic protocol that allows one party to securely transfer one of two messages to another party without revealing which message was chosen. Quantum oblivious transfer protocols use quantum entanglement to provide stronger security guarantees than classical protocols. Despite the promising security features of quantum cryptography protocols, practical implementations face challenges related to the limitations of current quantum hardware, the transmission distance of quantum channels, and the effects of noise and environmental factors. Quantum key distribution systems typically require specialized hardware for generating, transmitting, and detecting quantum states, making them susceptible to various technical limitations. The distance over which quantum entanglement can be maintained in a quantum channel is also limited, which can pose challenges for long-distance secure communication. Moreover, noise and imperfections in quantum hardware can introduce errors in the quantum states exchanged between the parties, requiring error correction and mitigation techniques. Quantum cryptography researchers are actively working on developing more robust and practical quantum key distribution systems and protocols that can operate in real-world conditions. One approach involves the use of trusted nodes or quantum repeaters to extend the range of secure quantum communication. Trusted nodes can act as intermediaries that help distribute quantum keys between distant parties, while quantum repeaters can extend the range of entanglement in quantum channels. Additionally, advances in quantum error correction codes and quantum hardware technology are expected to improve the reliability and performance of quantum cryptography systems. In summary, advanced quantum cryptography protocols harness the principles of quantum mechanics to provide unprecedented levels of security for key distribution, secure communication, coin tossing, and oblivious transfer. These protocols, such as the BBM92 and E91 protocols, rely on the properties of quantum entanglement and the no-cloning theorem to protect against eavesdropping. While practical challenges exist, ongoing research and technological advancements are paving the way for the deployment of secure quantum communication systems in the future. Quantum cryptography represents a promising avenue for addressing the security vulnerabilities posed by powerful quantum computers, ensuring the confidentiality and integrity of sensitive information in an increasingly connected world.

Secure quantum communication, often referred to as quantum cryptography, represents a cutting-edge field of study at the intersection of quantum physics and information theory. Quantum communication aims to achieve the highest level of security by leveraging the unique properties of quantum mechanics. Experts in this field are exploring innovative techniques and protocols to ensure the confidentiality and integrity of information exchanged between distant parties. One of the fundamental concepts in secure quantum communication is quantum key distribution (QKD). QKD enables two parties, often referred to as Alice and Bob, to establish a secret encryption key that is secure against any eavesdropping attempts by an adversary, typically denoted as Eve. The security of QKD relies on the principles of quantum mechanics, particularly the no-cloning theorem and the uncertainty principle. In a QKD protocol like the BB84 protocol, Alice generates a series of single photons, each prepared in one of two orthogonal polarization states, *such as horizontal or vertical. She sends these photons to Bob over a quantum channel. Bob randomly chooses a measurement basis (e.g., horizontal/vertical or diagonal/anti-diagonal) for each received photon and records the measurement outcomes. Alice and Bob then communicate openly over a classical channel to reveal the basis choices they used for each photon but not the measurement outcomes. They discard the measurement outcomes corresponding to different bases and keep those that match. The remaining outcomes form their secret key, which can be used for secure communication. The no-cloning theorem ensures that Eve cannot make perfect copies of the transmitted photons without disturbing them, making any eavesdropping attempts detectable. Quantum key distribution protocols like BB84 provide a provably secure way to establish encryption keys, even when the quantum channel is partially controlled by an adversary. Another key concept in secure quantum communication is entanglement-based quantum key distribution. Entanglement-based QKD protocols, such as the E91 protocol, rely on the creation and measurement of entangled particle pairs. Alice and Bob each have one particle from an entangled pair, and their measurement results on these particles determine their shared secret key. The advantage of entanglement-based protocols is that they allow secure key distribution even in the presence of fully cooperative adversaries. These protocols offer additional security against certain types of attacks. Quantum cryptography experts are not only focused on key distribution but also on the development of secure quantum communication channels for transmitting sensitive data. One such channel is the quantum teleportation protocol, which enables the transfer of an arbitrary quantum state from one location to another using quantum entanglement and classical communication. Quantum teleportation can be used* to establish secure quantum communication channels, as any interception or measurement of the transmitted quantum state would disrupt the entanglement and be detectable. Secure quantum communication for experts also involves addressing the challenges posed by practical implementations. Quantum key distribution systems require specialized hardware for generating, transmitting, and detecting quantum states. These systems are susceptible to various technical limitations, such as photon loss, noise, and environmental factors. To overcome these challenges, experts in the field are developing quantum repeaters and trusted-node networks. Quantum repeaters act as intermediaries that extend the range of secure quantum communication by creating entanglement between distant nodes. Trusted nodes are nodes in a network that can be relied upon for secure quantum communication, reducing the reliance on potentially untrusted intermediaries. In addition to hardware challenges, secure quantum communication experts are exploring new cryptographic protocols and techniques. These include device-independent QKD, which aims to ensure security without making assumptions about the quantum devices used in the protocol. Device-independent QKD protocols focus on verifying the security of quantum communication through violations of Bell inequalities. Experts are also investigating quantum communication protocols that can operate over long distances, potentially via satellite-based quantum links. Satellite-based quantum communication has the advantage of enabling secure communication between distant locations on Earth and providing a means to establish global-scale secure networks. Moreover, secure quantum communication experts are researching post-quantum cryptography. Post-quantum cryptography is a branch of cryptography that aims to develop cryptographic algorithms that remain secure even *in the presence of quantum computers. These algorithms are essential to protect data encrypted with classical methods from decryption by powerful quantum computers. Secure quantum communication experts are actively exploring the integration of post-quantum cryptographic algorithms into quantum key distribution and secure communication protocols. In summary, secure quantum communication for experts represents a dynamic and evolving field at the forefront of quantum technology and cryptography. Experts in this field are pushing the boundaries of quantum key distribution, exploring innovative protocols, addressing practical implementation challenges, and actively contributing to the development of post-quantum cryptography. Secure quantum communication is poised to play a pivotal role in safeguarding sensitive information in an era of increasingly powerful quantum computers, ensuring the confidentiality and integrity of data exchanged across quantum channels.*

Chapter 10: Interdisciplinary Perspectives on Quantum Computing

Quantum computing has emerged as a transformative technology with the potential to impact a wide range of multidisciplinary research areas. Its unique capabilities, driven by the principles of quantum mechanics, offer new avenues for solving complex problems that were once deemed intractable for classical computers. In recent years, researchers from diverse fields have recognized the value of quantum computing as a tool for advancing scientific knowledge and tackling real-world challenges. Quantum computing's ability to perform complex calculations at speeds unimaginable to classical computers is particularly appealing to disciplines that involve extensive computational modeling and simulation. One such area is quantum chemistry, where researchers seek to understand the behavior of molecules and chemical reactions at the quantum level. Quantum computers excel at simulating quantum systems, allowing for the accurate prediction of molecular structures and properties. This capability has the potential to revolutionize drug discovery, material design, and the study of chemical reactions, significantly accelerating the development of new pharmaceuticals and materials. In the field of materials science, quantum computing enables researchers to explore novel materials with tailored properties for various applications, such as superconductors for energy transmission and advanced materials for electronics. By simulating the behavior of atoms and electrons in these materials, quantum computers can guide the design of materials with improved characteristics and performance. Quantum computing's impact on materials science extends to the development of quantum materials themselves, which may lead to breakthroughs in quantum technologies and computing hardware. Multidisciplinary research involving quantum computing is not limited to the physical sciences. In the realm of finance and economics, quantum computing offers the potential to optimize complex financial models and perform rapid risk assessments. Quantum algorithms for portfolio optimization, option pricing, and risk analysis could enhance investment strategies and reduce financial market uncertainties. Furthermore, quantum computing can contribute to the optimization of supply chains, resource allocation, and logistics, benefiting various industries and improving efficiency in global trade. In the realm of healthcare, quantum computing holds promise for advancing personalized medicine and healthcare analytics. By analyzing vast amounts of patient data and genetic information, quantum algorithms can identify correlations, risk factors, and potential treatments for complex diseases. Drug discovery efforts can also benefit from quantum computing's ability to simulate molecular interactions, accelerating the search for new therapies. The field of artificial intelligence (AI) and machine learning is another area where quantum computing has the potential to make significant contributions. Quantum machine learning algorithms, such as quantum support vector machines and quantum neural networks, can enhance AI capabilities by solving optimization problems faster and more efficiently than classical counterparts. These algorithms can be applied to tasks like image recognition, natural language processing, and data analysis, opening new avenues for AI-driven advancements. Moreover, quantum computing enables the efficient factorization of large numbers, a crucial capability with implications for cryptography and cybersecurity. Researchers in computer science and cryptography are exploring quantum-resistant encryption schemes to protect data from future quantum threats. Quantum-safe cryptography is essential for securing sensitive information in an era where quantum computers could potentially break classical encryption methods. Quantum-resistant algorithms, such as lattice-based cryptography and hash-based cryptography, are being developed to safeguard digital communication and protect against quantum attacks. In the field of energy, quantum computing holds promise for optimizing the design and operation of renewable energy systems and improving energy storage technologies. By simulating the behavior of quantum materials and chemical reactions, quantum computers can contribute to the development of more efficient solar cells, batteries, and energy conversion devices. Quantum computing can also assist in the optimization of energy distribution and grid management, ensuring the stability and sustainability of future energy networks. Aerospace and aviation industries can benefit from quantum computing in various ways, including the optimization of aircraft design, flight routing, and air traffic management. Quantum algorithms can address complex optimization problems in these domains, leading to more fuel-efficient and environmentally friendly aviation solutions. Additionally, quantum computing's ability to analyze large datasets can enhance the safety and reliability of aircraft through predictive maintenance and fault detection. Environmental science and climate research stand to gain from quantum computing's computational power. Quantum simulations of complex climate models and environmental processes can provide more accurate predictions of climate change, weather patterns, and natural disasters. This knowledge is invaluable for policymaking, disaster preparedness, and sustainable environmental management. The fusion of *quantum computing and multidisciplinary research extends to the exploration of quantum algorithms for solving problems in fields as diverse as cryptography, finance, healthcare, AI, energy, aerospace, and environmental science. Researchers from these disciplines are collaborating to harness the potential of quantum computing to address complex challenges and advance knowledge across the scientific and technological spectrum. As quantum hardware continues to evolve and become more accessible, the integration of quantum computing into multidisciplinary research is expected to accelerate, leading to groundbreaking discoveries and innovative solutions to global challenges. Interdisciplinary collaboration is essential to fully exploit the capabilities of quantum computing and leverage its transformative potential for the betterment of society. In summary, quantum computing has emerged as a powerful tool for multidisciplinary research, offering new ways to tackle complex problems in diverse fields. From quantum chemistry to finance, healthcare, AI, energy, aerospace, and environmental science, quantum computing's computational advantages are driving innovation and advancing scientific knowledge. Interdisciplinary collaboration is key to unlocking the full potential of quantum computing and addressing the most pressing challenges of our time. Collaboration is a cornerstone of progress in quantum computing, a field rife with complex challenges. As the quantum computing landscape continues to evolve, researchers and experts recognize the importance of working together to overcome hurdles and unlock the full potential of this revolutionary technology. Quantum computing challenges span a wide spectrum, from hardware development and error correction to algorithm design and application development. Each of these domains presents unique obstacles that can be addressed more effectively through collaborative approaches. Quantum hardware development, for instance, is a field where cooperation among academic researchers, industry experts, and government institutions is crucial. The fabrication of qubits, the building blocks of quantum computers, requires cutting-edge technologies and substantial resources. By pooling resources and expertise, collaborations can accelerate the development of reliable and scalable quantum hardware. Furthermore, collaborative efforts can facilitate the sharing of knowledge and best practices, ensuring that advancements in quantum hardware benefit the broader research community. Quantum error correction is another area where collaboration is pivotal. Quantum computers are highly susceptible to errors caused by decoherence and environmental noise. Developing error-correcting codes that can mitigate these errors is a complex task that requires the collective efforts of researchers from various disciplines. Collaborative projects can drive the development of robust error correction techniques, making quantum computation more reliable and accurate. Moreover, as quantum hardware continues to evolve, new error models and mitigation strategies will need to be developed collaboratively to address emerging challenges. Algorithm design in quantum computing is yet another domain where collaboration is essential. Quantum algorithms have the potential to solve problems exponentially faster than classical algorithms, but devising these algorithms can be highly intricate. Collaborative efforts between computer scientists, mathematicians, and physicists can lead to the discovery of novel quantum algorithms that tackle critical problems in fields like cryptography, materials science, and optimization. Collaboration in quantum algorithm development also extends to open-source quantum software development.*

Quantum software frameworks like Qiskit, Cirq, and Forest are developed collaboratively by organizations and individuals. These open-source projects allow researchers and developers worldwide to access quantum computing resources and contribute to the development of quantum algorithms and applications. Moreover, quantum algorithm designers often collaborate with experimentalists to test and validate their algorithms on real quantum hardware, providing valuable feedback for improvement. Application development in quantum computing is an area that benefits significantly from interdisciplinary collaboration. Researchers from fields such as chemistry, finance, machine learning, and optimization are working alongside quantum computing experts to explore the practical applications of quantum algorithms. For example, in the field of quantum chemistry, collaboration between quantum algorithm designers and computational chemists has led to the development of quantum algorithms that can accurately simulate molecular systems. This collaboration has the potential to revolutionize drug discovery, materials science, and the understanding of chemical reactions. In finance, quantum algorithms for portfolio optimization and risk assessment are being developed collaboratively to enhance investment strategies and reduce financial market uncertainties. Machine learning researchers are also collaborating with quantum computing experts to harness the power of quantum algorithms for tasks such as data analysis, pattern recognition, and optimization. Interdisciplinary collaboration is essential in addressing the scalability challenges of quantum computing. As quantum computers grow in size and complexity, the need for scalable quantum algorithms and software becomes paramount. Collaborations between quantum hardware developers, quantum algorithm designers, and software engineers can lead to the creation of efficient and scalable quantum computing platforms. Furthermore, collaborations between academia, industry, and government agencies can help establish a robust quantum ecosystem that supports the growth of quantum computing. This includes the development of quantum education and training programs to prepare the next generation of quantum scientists and engineers. Collaboration in quantum computing extends to the international level, with countries around the world recognizing the strategic importance of quantum technology. International collaboration on quantum research fosters the sharing of knowledge, resources, and expertise. It also promotes the development of common standards and protocols that facilitate the interoperability of quantum hardware and software across borders. Moreover, international collaboration helps ensure that the benefits of quantum computing are equitably distributed, with a focus on addressing global challenges, such as climate change and healthcare. Government agencies, research institutions, and private companies are forming international partnerships to accelerate progress in quantum computing and ensure its responsible development. Ethical considerations also play a role in collaborative approaches to quantum computing. As quantum computing advances, ethical questions surrounding data privacy, security, and the potential misuse of quantum technology must be addressed collaboratively. Researchers, policymakers, and ethicists are working together to establish ethical frameworks and guidelines that promote the responsible use of quantum computing. Collaborative efforts in quantum computing are not limited to academia and industry; they also involve quantum communities, hackathons, and competitions. These events provide a platform for researchers, students, and quantum enthusiasts to collaborate on quantum projects, exchange ideas, and showcase innovations. Collaborative initiatives like the

Quantum Open Source Foundation (QOSF) and the Quantum Computing Collaboration (QCC) are dedicated to fostering collaboration and open-source development in quantum computing. Quantum hackathons and quantum competitions encourage participants to work together on solving quantum computing challenges and exploring new applications. In summary, collaborative approaches are fundamental to addressing the multifaceted challenges of quantum computing. By bringing together experts from diverse fields, fostering international cooperation, and promoting ethical considerations, collaborations pave the way for breakthroughs in quantum hardware, error correction, algorithm design, and application development. Quantum computing's potential to revolutionize industries and solve complex global problems can be fully realized through interdisciplinary and international collaboration, making it a truly transformative technology for the benefit of society.

Conclusion

In the realm of quantum computing, the journey from curiosity to mastery is one of exploration, discovery, and boundless potential. As we conclude this book bundle, "Quantum Computing: Computer Science, Physics, and Mathematics," we reflect on the incredible journey we've embarked upon.

In "Quantum Computing Demystified: A Beginner's Guide," we introduced the fascinating world of quantum computing to newcomers, laying the foundation for understanding its core principles and potential applications. With each page, readers took their first steps into a realm where qubits, superposition, and entanglement reign supreme.

"Mastering Quantum Computing: A Comprehensive Guide for Intermediate Learners" took us deeper into the quantum landscape, where we delved into advanced topics, quantum algorithms, and quantum programming. Intermediate learners honed their skills, becoming proficient in harnessing the quantum power that underlies this remarkable technology.

"Advanced Quantum Computing: Exploring the Frontiers of Computer Science, Physics, and Mathematics" was our gateway to the cutting edge. In this book, we ventured into uncharted territory, exploring the forefront of quantum computing, including quantum error correction, quantum cryptography, and quantum simulations. Experts and enthusiasts alike were challenged and inspired by the possibilities on the horizon.

Lastly, in "Quantum Computing: A Multidisciplinary Approach for Experts," we discovered that quantum computing transcends disciplines. Together, we explored its role in computer science, physics, and mathematics, recognizing its potential to reshape industries and address global challenges. Quantum computing *became more than a technological endeavor; it became a collaborative force for innovation and transformation.*

As we bid farewell to this book bundle, we invite you to continue your quantum journey. Quantum computing is not just a field of study; it's a realm of endless exploration. It's a bridge between the known and the unknown, offering solutions to problems that once seemed insurmountable.

We hope that the knowledge gained from these books will empower you to engage with the quantum world, contribute to its advancement, and embrace the opportunities it presents. Whether you are a beginner taking your first steps or an expert pushing the boundaries of what's possible, quantum computing offers a path to understanding the universe at its most fundamental level.

In this multidisciplinary bundle, we've unveiled the mysteries of quantum computing, peered into its potential applications, and navigated the complexities of this transformative technology. As you close these books, remember that the quantum journey never truly ends; it evolves with each discovery and collaboration.

We encourage you to continue your exploration, engage in the quantum community, and stay curious. The future of quantum computing is shaped by those who dare to dream, those who challenge the limits of classical computing, and those who understand that the quantum realm holds the keys to unlocking a new era of knowledge and innovation.

Thank you for joining us on this quantum odyssey. May your quantum adventures be filled with wonder, discovery, and the pursuit of answers to the universe's most profound questions. The quantum frontier awaits, and the possibilities are boundless.